Futures of Artificial Intelligence

Futures of Artificial Intelligence

Perspectives from India and the U.S.

ROBERT M GERACI

Professor of Religious Studies at
Manhattan College in New York City

OXFORD
UNIVERSITY PRESS

OXFORD
UNIVERSITY PRESS

Oxford University Press is a department of the University of Oxford.
It furthers the University's objective of excellence in research, scholarship,
and education by publishing worldwide. Oxford is a registered trade mark of
Oxford University Press in the UK and in certain other countries

Published in India by

Oxford University Press

22 Workspace, 2nd Floor, 1/22 Asaf Ali Road, New Delhi 110 002, India

ISBN-13 (print edition): 978–8–19–483167–9

ISBN-10 (print edition): 8–19–483167–9

ISBN-13 (eBook): 978–9–39–105030–6

ISBN-10 (eBook): 9–39–105030–1

ISBN-13 (OSO): 978–9–39–105032–0

ISBN-10(OSO): 9–39–105032–8

DOI: 10.1093/oso/9788194831679.001.0001

Typeset in Minion Pro 10.5/14
by Newgen KnowledgeWorks Pvt. Ltd., Chennai, India
Printed in India by Rakmo Press Pvt. Ltd.

This book is dedicated to my brilliant, loving, and beautiful wife, Jovi.

May there be no end to the wonders we experience together.

Contents

Acknowledgements

I wrote this book at a time when I really should have been working on another book, one that I am writing in collaboration with my wife, Jovi, and for which Jovi and my children—Zion and Lily—accompanied me on a return trip to India. So, of course, I must thank Jovi for her patience and my children for their willingness to travel the world with me. They are all the family I could ever hope for.

My friends and colleagues in India were crucial to the development of this book, and a few deserve special mention. Dr Renny Thomas was a sounding board to complaints, always ready to provide me with a long list of additional reading, and an unending font of encouragement. His thoughts on religion and science and on Indian scientific communities are at the forefront of our field. Ravi Menezes, the owner of Goobe's Book Store in Bangalore, jumped around town with me, hosting me in several venues so that I could share some of these ideas in the city's tech community. As icing on the cake, I even got to bang my head on the ceiling of a cave temple we visited after one of those events. Ravi is an indefatigable advocate for reading, learning, an environmentally friendly urban future, and practically everything else worthwhile—I'm grateful for his excitement over my work and efforts to share it in Bangalore. The Indian Institute of Science and the National Institute of Advanced Studies gave me intellectual homes in 2012–2013 and 2018–2019, respectively. I am grateful to IISc and its Centre for Contemporary Studies (now the Centre for Science and Policy) and to NIAS for hosting me and giving me the opportunity to learn about scientific cultures in India, as well as to all the scientists and engineers who've shared their time with me over the years.

Shiny Das and then Nandini Ganguli at OUP India were enthusiastic from the start of our conversations about publishing this book. Nandini worked with me through transitions, delays, and the chaos of COVID-19; her ongoing encouragement and engagement with my work provided support and relief through trying times. Rajakumari and Ayshwarya at OUP were terrific as we handled all the technical processes; I appreciate their swift and precise work in helping this book come to print.

Cynthia Read at OUP deserves my profound gratitude for support and intervention that twice preserved this project when our chaotic times produced administrative complications. She is also among the most brilliant conversationalists I've had the privilege to lunch with, and her insight into the field is unparalleled.

In New York, I have colleagues who unceasingly support my research and teaching labors, no matter how unconventional they may be. I'm lucky to have so many terrific colleagues. Dr Cory Blad and Dr Kevin Ahern, in particular, have provided the perfect mix of intellectual partners in crime and committed friendship—they have become brothers. Dr Stephen Kaplan has long been a mentor and is now collaborator as we contemplate the connections between Hinduism and AI. Members of the Manhattan College staff make doing my job much easier: Brendon Ford was amazing in helping me acquire materials from other libraries, engaging with gusto even when I gave him rare and nettlesome requests; Dianna Cruz, Angie Thrapsimis, Syrita Newman, MaryEllen LaMonica, and Jamie McGuinness helped me with so many tasks, both large and small, that I'd still be running around in circles without them.

One's academic community can make research labors remarkably fun. Dr Leslie Smith, Dr Cat Newell, and Dr Alex Ornella make conferences twice as fun to attend. Dr Peter Boothe may have left Manhattan College, but he read this book and his conversation afterward was insightful. Mr Samanth Subramanian was kind enough to respond to my shot-in-the-dark emails, consulting his own notes on Jack Haldane and even emailing a friend to trace historical connections that would have entirely eluded me otherwise. Philip Deslippe was absolutely invaluable: it was thanks to his effort that I was able to access the mid-20th century volumes of *Indian Rationalist*. Dr Rudy Busto—for years a brilliant friend and academic conspirator—was kind enough to introduce us, and so without him, I would still be wondering how to get where I needed to be. All these folks make work more joyful through their help, encouragement, and friendship.

I am grateful that the arrangement of these ideas produced new opportunities to share and learn in international contexts. Professor Henning Glaser invited me to share some themes from this book at the German Southeast Asian Center of Excellence for Public Policy and Good Governance in Bangkok, Thailand. Dr Pasquale Annichhino invited me to do likewise at the Center for Religious Studies at the Bruno Kessler

Foundation (FBK) in Trento, Italy. These visits came as I was finishing the manuscript and I value the collaborations they produced, which include ongoing work with Dr Marco Ventura and Dr Boris Rähme at FBK and, through Boris, with Dr Inken Prohl of Heidelberg University.

The main chapters of this book were composed out of lectures given to audiences in the U.S. and Europe under the auspices of various keynote lectures. First, I am grateful to Dr Michael Greenwald and the faculty of St. Lawrence University in New York for inviting me to deliver the 2016 Kathryn Fraser Mackay Memorial Lecture and to the Mackay family for their sponsorship of the event. Not only did I enjoy visiting the campus and my daylong conversation with Michael, whose intellect is as keen as his spirit is generous, but I even got a liberal dose of amusing anecdotes about past lectures. Second, Dr Alexander Darius Ornella invited me to give the closing keynote lecture at the Apocalypse and Authenticity conference of the International Theology, Religion and Popular Culture Network, hosted at the University of Hull. Alex is a treasured friend and colleague who is clearly the equal of his namesakes. Third, Simon Robinson brought me back to the UK to offer the opening keynote to the AI and Apocalypse Symposium hosted by the Centre for the Critical Study of Apocalyptic and Millenarian Movements. Speaking about apocalypticism at the home of the Panacea Society, a 19th–20th-century millenarian group anticipating global transformation was keenly appropriate. Further, Simon joined me in collecting essays from a few of the conference speakers and publishing them as a symposium in *Zygon: Journal of Religion and Science*. *Zygon*'s then outgoing editor, Dr Wim Drees, and its incoming editor, Dr Arthur Peterson, were enthusiastic in bringing the project to print. Finally, I thank Dr Karoline Krenn for bringing me to Berlin as keynote speaker for Unfathomable: AI as Founder of Order, a conference hosted by Fraunhofer FOKUS. That venue and its participants were truly memorable—though I regret that I could not follow the German speakers. Parts of Chapter 4 (translated to German) were published by Fraunhofer FOKUS as 'Rekodierung von Religion: Theologische Darstellungen von Künstlicher Intelligenz und der Zukunft von Gesellschaft' in the volume *(Un)ergründlich? Künstliche Intelligenz als Ordnungsstifterin*.

As the first people and the last people I think of each day, Jovi, Zion, and Lily are not just the first to be acknowledged here but the last as well. Zion and Lily, now teenagers, will happily discuss whether we're living in a computer simulation or whether life would be more fun fighting against (or better yet, alongside) dragons. In short, they've become intellectual partners even as they remain my favourite playmates. Jovi's contributions to my intellectual and emotional life are too numerous to detail, but I hope she knows how much I appreciate her support and how much I look forward to finishing that other book with her now that this one is finished.

About the author

Robert M Geraci earned his PhD from the University of California at Santa Barbara and is currently Professor of Religious Studies at Manhattan College in New York City. He has been a visiting researcher at the Carnegie Mellon University Robotics Institute and has been Fulbright-Nehru Research Scholar at the Indian Institute of Science (2012–2013) and at the National Institute of Advanced Studies (2018–2019), both in Bangalore. He is the author of *Apocalyptic AI: Visions of Heaven in Robotics, Artificial Intelligence, and Virtual Reality* (Oxford 2010), *Virtually Sacred: Myth and Meaning in World of Warcraft and Second Life* (Oxford 2014), and *Temples of Modernity: Nationalism, Hinduism, and Transhumanism in South Indian Science* (Lexington 2018). His research has been supported by the U.S. National Science Foundation, the American Academy of Religion, and two separate Fulbright-Nehru Awards. He is a Fellow of the International Society for Science and Religion.

Introduction

A beginning to the end

For at least two thousand years, religious practitioners have been an-
nouncing the imminent end of the world. Today, human-driven climate
change and radical technological developments combine to make such
an end appear plausible to even the most staunchly atheist observer. Few
believe that such catastrophes would be divinely ordained, but the fear
of world-ending cataclysm looms and suggests new religious responses.
Whether the end of life as we know it will bring extinction to the human
race (and perhaps all life on Earth) or is a prelude to a radical and wondrous new world remains, at this point, an open question. How one views
such ends is, of course, deeply inflected by religious culture. This book
examines how two religious cultures perceive the end of the world as
threatened or promised by the complex scientific networks of artificial
intelligence (AI).

AI is now central to questions about the destiny of humanity and the
world, including fears of our imminent demise. Twentieth-century scientists established the Doomsday Clock as a measurement of humanity's
progress towards extinction, a decision motivated by the development of
atomic warfare. In the 21st century, fewer commentators fear that atomic
weapons threaten the total extinction of humanity; and yet non-nuclear
threats mean that the clock remains close to midnight. Political parti-
sanship, nationalist prejudice, and environmental threats are among the
key reasons for this fear, but it takes little imagination to picture a world
where AI becomes the deciding factor in the clock's advance.

New computing technologies offer reason to believe that humanity will
witness extraordinary progress in AI, progress that would surely reconfigure the world if it leads to human-equivalent or greater-than-human
machine intelligence. Such AIs might work alongside humanity and

Futures of Artificial Intelligence. Robert M Geraci, Oxford University Press. © Oxford University Press 2022.
DOI: 10.1093/oso/9788194831679.003.0001

resolve the Earth's environmental crisis, cleaning up pollution in the land, water, and air, protecting endangered species, and offering new modes of industrial production. Alternately, they might decide that humanity is an irretrievable plague upon the Earth and determine that we have out-lived our evolutionary use. More sober predictions tend to fall between paradise and hell, suggesting that our developing technologies will assist us in solving some of our problems but surely will fall short of creating a perfect world. Likely, they will assist us in important ways and then add unexpected difficulties. Moderates claim that while AI is unlikely to spell the doom of humanity, it could easily exacerbate existing inequalities in the distribution of power. In keeping with these latter possibilities, the *Bulletin of the Atomic Scientists*, which maintains the Doomsday Clock, published an essay in 2019 decrying AI-fueled surveillance and the pos-sibility of government control.[1] Many scientists also worry about the out-comes if large numbers of robots are equipped with weapons.[2] These and other fears challenge us to seek novel approaches in AI.

Without careful design, the dangers posed by AI will drown us. We already have evidence that poorly constructed AI systems lead to un-expected problems. In the banking industry, we see that AI exacerbates structural injustices through race-based loans.[3] In judicial decision-making, AI reifies prejudice and ensures harsher outcomes for people of colour.[4] Although it might seem less harmful that facial recognition does not efficiently identify darker skin tones, the fact that driverless cars have more difficulty spotting non-white individuals shows there is literal risk to life and limb that accompanies poorly trained pattern recognition.[5] Meanwhile, the deployment of military robotics surely comes with some

---

[1] Sherman, 'Digital Authoritarianism and the Threat to Global Democracy'.
[2] For example, Yoshua Bengio, a noted AI researcher; see Bilefsky, 'He Helped Create A.I. Now, He Worries about "Killer Robots"'. Theologians have also entered this sphere, pressing for a ban on military robots (Green, *Robots and AI*, 216–19).
[3] Bartlett, et al., 'Consumer-Lending Discrimination in the FinTech Era'.
[4] Angwin, et al., 'Machine Bias'; Mayson, 'Bias In, Bias Out'.
[5] Porter, 'Federal Study of Top Facial Recognition Algorithms Finds "Empirical Evidence" of Bias'; Wilson, Hoffman, and Morgenstern, 'Predictive Inequity in Object Detection'. The prob-lems with such technology, especially as used by law enforcement, led to IBM's high-profile de-parture from the facial recognition research and marketplace; see Peters, 'IBM Will No Longer Offer, Develop, or Research Facial Recognition Technology'. Lepri, et al. note the importance of identifying clear mechanisms of accountability, fairness, and transparency in algorithmic decision-making through data protection, vetting of algorithms, and auditing decisions ('Fair, Transparent, and Accountable Algorithmic Decision-making Processes').

humanitarian benefits but could accelerate the willingness of wealthy nations to conduct asymmetric warfare. The race to develop increasingly powerful military AI seems hardly conducive to a healthy human future. Of course, all these risks are the obverse of the more optimistic aspirations of AI researchers to create fairer, more efficient decision-making. Despite the potential for good, however, AI poses legitimate and strong risks to human flourishing.

New religious perspectives, emergent from tech enthusiasm, offer some respite from this gloom. In fact, many tech visionaries see AI offering redemption from the very problems we face. Most traditional religions are already attuned to redemptive perspectives and to positive futures, so their potential use in this way should come as no surprise. But the quasi-religious perspectives being built into our technologies often hide in the background, and thus the direction of their impact is cloudy. While religious value systems come with their own potential failure modes, they might be fruitfully engaged in the global development of AI. With proper attention to the value systems that already exist, and that are often already incorporated into our cultures, we might leverage them as we develop policies and practices for digital technologies.

Scientific and technological outcomes are dependent upon human decisions, and these are themselves dependent on, though not completely determined by, the cultural landscape of the decision-makers. This book reveals cultural threads entwined with AI that can be leveraged for good or ill. It is my hope that humanity can make global progress towards a better future, but for this to come true, AI technologies must benefit from global contributions. And so in the chapters that follow, this book uncovers some, but not all, of the cultural realities that shape how people see and use AI technologies. It further indicates what cultural resources might be deployed to help us create a better future.

Cultures of technology

We often consider science and technology to be value-free and culture-neutral. In important respects, such claims are true. For example, mathematics works to describe the natural world whether one is in Bangalore or Austin. Regardless of place or time, mathematics is the most precise language employed by humanity, and most sciences leverage this precision. However, there are important ways in which science and technology can never be value-free or culture-neutral. People use technologies to shape the cultures in which they live, but their cultures impact how they look upon and employ technologies. Thinking about science, technology, and culture sometimes requires that we differentiate one culture from another (perhaps using tricky terms such as 'western' and 'nonwestern'), a task that can be quite difficult. I will return to that difficulty in the conclusion of this book; but at the outset, I note that cultures are never monolithic, never separated from one another, and their demarcation is always a political as well as intellectual effort. After all, we choose to identify some phenomena as relevant to one culture and not others. Such choices also characterize our understanding of science and technology. That is, one can imagine science being entirely free from the cultures that ensconce it but, as this book will show, there are many ways in which culture is tied to AI. Exploring those cultures matters for ensuring the humane use of AI technologies.

National and international groups have begun building an ethical and practical framework for the development of AI. Much of this work gets naturally enfolded into military ethics,[6] but more wide-ranging policy documents are already adding to the conversations. For example, the European Union produced a set of ethical guidelines for AI in 2019, though the commission was criticized for not including enough philosophers and ethicists.[7] The Japanese government chose to situate AI technologies within a larger perspective on technology, and thus their earliest effort lacks the detail of the European Union's draft guidelines. Japan released its 2016 science and technology plan with a clear interest in moral outcomes of their deployment of technology: the policy document refers to the nation as one that 'constantly contributes to the advancement of mankind'.[8] This global perspective needs to be more widely shared by the world's governments and their approaches to science and technology.

---

[6] For example, Singer, *Wired for War*; Gill, 'Artificial Intelligence and International Security'.
[7] Metzinger, 'EU Guidelines: Ethics Washing Made in Europe'. Given that the EU's group entirely lacked people who understand the cultural roots of our technologies, philosophers might content themselves with having at least some representation.
[8] Government of Japan, 'The 5th Science and Technology Basic Plan', 7.

The official Japanese position favours social transformation and, in this sense, mirrors American approaches described throughout this book.[9] But social and global transformation takes many forms: in contrast to their popularity in the U.S., there appears to be little enthusiasm for dreams of AI transcendence or transhuman futures in Japanese robotics.[10] There are, of course, other differences between cultural approaches to robotics. For example, Jennifer Robertson notes that the Japanese government's *Innovation 25* proposal 'gives priority to public order over individual sovereignty and freedom.'[11] This may represent a difference between Japanese cultural values and public policy compared to other governmental approaches. While neither Japanese nor American perspectives are homogenously positive or negative about AI, Japanese expectations about the future tend towards a collaborative vision of humanity alongside intelligent machines, whereas Americans often fear the latter replacing the former (though such fear certainly has less purchase in everyday practice than it does in Hollywood movies).[12] As these positions are not independent of religious influence, it reveals the importance of understanding what cultural resources can be deployed throughout the world in the advancement of global human interests.

Like other nations, India has taken steps towards a policy framework for AI. This is eminently desirable given the remarkable growth of AI research in India during the 21st century.[13] Early commentators have noted

---

[9] Government of Japan, 'The 5th Science and Technology Basic Plan', 8. For an interpretation of how the Japanese plan intersects with and possibly produces new visions of transhumanism, see Gladden, 'Who Will Be the Members of Society 5.0?'

[10] Robertson, *Robo Sapiens Japanicus*, 3, 174.

[11] Robertson, *Robo Sapiens Japanicus*, 36.

[12] See, especially, Geraci, 'Spiritual Robots' but also Geraci, *Apocalyptic AI*, 44, 54, 106, 142, 150. Stated Japanese goals typically favour co-existence and cooperation among human beings and machines; see Government of Japan, 'The 5th Science and Technology Basic Plan', 13. Jennifer Robertson describes such coexistence as broadly constitutive of Japanese discourse on robotics; see *Robo Sapiens Japanicus*, 190. For a brief mention of how positive and negative perceptions appear in both Japan and the U.S. as a criticism of my own work, see Rambelli, 'Dharma Devices, Non-Hermeneutical Libraries, and Robot-Monks', 67; Gould and Walters, 'Bad Buddhists, Good Robots', 285. Rambelli's perspective, if not also that of Gould and Walters, makes overmuch of what he sees as my belief that Japanese perspectives are always positive about robotics and American perspectives negative. Rambelli mistakes my argument that Japanese religious perspectives contribute to an emphasis upon robotics as opposed to AI for a claim that Japanese religions lead the people to universally love robots. This confusion is somewhat surprising given Rambelli's skillful argument that Japanese Buddhism has a long tradition of utilizing and emphasizing machines—an argument that actually supports my own in 'Spiritual Robots'.

[13] On the publication metrics of Indian AI researchers, see Shrivastava and Mahajan, 'Artificial Intelligence Research in India'. Their analysis shows a stratospheric rise in Indian publications

the necessity of such efforts and have presented them in the context of human-equivalent machine intelligence. For example, one law professor notes the importance of establishing rules that govern a machine's legal personhood and expectations of its intentionality.[14] NITI Aayog, an official policy think tank of the Indian government, released its 'National Strategy for Artificial Intelligence' in 2018. The document advocates 'AIforAll', its branding for 'inclusive technology leadership, where the full potential of AI is realized in pursuance of the country's unique needs and aspirations'.[15] A similar ethos can underwrite the global approach to AI, and this book argues that our cultural values *about* AI would benefit from the same approach.

India and the U.S. have separate cultural resources that can play a role in how scientists, policymakers, and the public exploit technological progress. The rise and role of transhumanism in each nation makes some of these cultural resources explicit, particularly their religious influences. Transhumanism is a philosophical and/or religious movement that articulates visions of the future; it is a movement that advocates human evolution and transcendence over biological limits through the use of science and technology. There is a meaningful sense in which all people are transhumanists: we use technology to overcome our limitations (e.g. eyeglasses to help us read[16]). But avowed transhumanists seek a higher goal than alleviating common and basic distress. Their pursuit goes beyond therapy and into the realm of augmentation; their intellectual labours could lead towards a posthuman species through genetic engineering, cyborg integrations, or transferring human minds into machines. While further reflections on transhumanism are to come in future chapters, it is important to notice here how the pursuit of human augmentation leads naturally to reflections on the end: the end of humanity, the end of biology, the end of earthly life. But transhumanism leads similarly to reflections on what might come after: a new species, a new world.

---

on AI topics but tempers this by showing a dramatic reduction of Indian scientists' impact in the field (as measured by citations).

[14] Bajpai and Irshad, 'Artificial Intelligence, the Law and the Future'.
[15] NITI Aayog, 'National Strategy for Artificial Intelligence', 7.
[16] In fact, even writing can be considered a transhumanist technology for overcoming individual and social memory dysfunction.

So the flow of transhumanist ideas provides the perfect illustration of how culture matters in the future of technology. Transhumanism is, itself, a cultural resource, now ready-to-hand for humanity poised at the brink. Accepting, rejecting, and more often modifying transhumanist beliefs about technology, people in both India and 'the West'[17] see opportunities not only to absorb technologies into their lives but also to infuse technologies with their own local modes of living. AI is bound to transhumanism, having been born in the same era and offering much of the technological structure that supports transhumanist faith.

Although transhumanism challenges us to evaluate our expectations for technology, I will maintain a pretense towards objectivity: this book does not judge the truth value or likelihood of the central transhumanist positions described. What I call Apocalyptic AI—the belief that human beings will transcend their limits by merging with machines in a glorious new post-biological world to come—runs rampant in 21st-century conversations about technology and circulates through global visions of AI. Both advocates and critics of these apocalyptic perspectives find themselves central to public debates about AI. Instead of judging the truth of such beliefs, this book traces the cultural configurations of Apocalyptic AI and the dynamics of their transmission. It investigates the way that Apocalyptic AI operates in global science, considering its passage into a different culture, and argues that the contributions of other cultures could fruitfully be placed alongside the apocalyptic perspectives common to western technological imaginaries when we contemplate the future. We ought to look for the ways that differing communities see science and technology so that we can leverage the best contributions of each.[18]

At its core, this book suggests that in order to produce worthwhile outcomes, we should begin by analysing what perspectives, including Apocalyptic AI, are already built into science and technology. Among these, there are religious perspectives that often get lost in the vigorous debates over scientific politics or the social practice of science. While the history of science clearly reveals the significance of religious ideas in the rise of modern science, we much less frequently see the ways in which

---

[17] A brief evaluation of such terms will take place in the conclusion to the book.
[18] I have endeavoured to trace a few such ideas in collaboration with Yong Sup Song; see 'Global Culture for Global Technology'. On the concept of 'technological imaginary', see Ornella, 'Towards a "Circuit of Technological Imaginaries"'.

religion either advances or prohibits scientific discovery in the contemporary world. Indeed, some people say that religion *only* prohibits scientific discovery; but such claims hinge upon an impoverished understanding of religion: that religion is only institutional religion, that it manifests only as conservative institutions, and that it favours dogmatic expressions of belief over pragmatic interest in empirical reality. But there are other kinds of religion and many ways religion engages with scientific and technological practice.

Scholarship in the study of religion and science long ago debunked the idea that religion and science are in a state of necessary conflict. Certainly, some people fight with other people over religious or scientific issues, but many scholars (some of whom will be featured later in this book) have conclusively shown that matters are far more complex than any simple narrative of conflict can describe.[19] Elsewhere, I have argued that one key to future research is moving beyond single-minded attention to religious beliefs and scientific theories.[20] It is the overwhelming attention to people's beliefs (especially as indicated by religious scriptures) that has led to the easy adoption of the conflict thesis. Despite the actual behaviour of people, some of our contemporaries persist in the claim that animosity reigns between religion and science. But in reality—in their professional lives or in their public performances—people engage with religion and science in a variety of ways; some of these interactions involve intellectual or political conflict, but many involve other dynamics. Scholarly inquiry into the relationship between religion and AI, while still needing further development, clearly demonstrates some of these.

Investigations into the relationships between religion and AI began decades ago, though it is only in the second decade of the 21st century that a more general public, both academic and civic, attended to these questions. There are classic inquiries into the ethical significance of AI, such as cyberneticist Norbert Wiener's *God & Golem: A Comment on Certain Points Where Cybernetics Impinges on Religion* (1964), and occasionally a fascinating collection that illustrates the panoply of cultural

---

[19] The most famous narrative of conflict is White, *The History of the Warfare of Science with Theology in Christendom*. Although this narrative has met with many criticisms, too many to enumerate here, the most famous effort to categorize religion and science relationships into a typology that includes other options is that of Ian Barbour; see Barbour, *Religion and Science*.
[20] Geraci, *Temples of Modernity*, 173–8; Geraci, 'A Hydra-logical Approach'.

responses to AI, such as Harry Geduld's and Ronald Gottesman's *Robots, Robots, Robots* (1978). Although Edmund Furse tried to launch a debate about religion and robotics in the 1980s, the study of AI and religion by people dedicated to the study or practice of religion did not really begin until the 1990s and gained sophistication in the early 2000s.[21] Seminal scholars include Noreen Herzfeld, Anne Foerst, and Antje Jackelén.[22] By the end of a decade, more scholars joined the field, including some who considered the ramifications of AI from outside the Christian fold.[23] Simultaneously, other scholars began finding correspondences between the operations of religious practice and the human response to robots.[24]

Thanks in large part to experimental collaborations between religious groups and robotics developers, something of a cottage industry in the study of robotics and religion erupted by the close of the second decade of the 21st century. Early 'monk robots' had been nothing more sophisticated than audio playback devices, but later projects utilized more sophisticated AI. The BlessU-2 robot tested in Germany, the Nissei Eco robot built to perform Buddhist funerals in Japan, and the robot arm used to make offerings to Ganesha in India were all media darlings in 2017 and prove that human beings willingly experiment with AI in religious settings.[25] With robots engaged in religious work and human beings increasingly connected to quasi-intelligent AI assistants (e.g. Siri and Alexa), scholars enthusiastically look towards the intersections of AI and religion. For example, scholars suggest that AI could be used to generate sermons or offer pastoral counselling, that AI combined with augmented reality could re-enchant the world, and that robots might borrow

---

[21] Furse, 'The Theology of Robots'; Furse, 'Towards the First Catholic Robot?' For a dismissive response to Furse, see Barker, 'Edmund Furse's "The Theology of Robots"'.

[22] Foerst, 'Cog, a Humanoid Robot, and the Question of the Image of God'; Foerst, *God in the Machine*; Herzfeld, 'Creating in Our Own Image'; Herzfeld, 'Cybernetic Immortality versus Christian Resurrection'; Jackelén, 'The Image of God as *Techno Sapiens*'.

[23] For example, Tamatea, 'Online Buddhist and Christian Responses to Artificial Intelligence'; Geraci, 'Spiritual Robots'. More recently, Kaunda synthesizes Christian theology, AI, and non-Christian religious ideas in 'Bemba Mystico-Relationality and the Possibility of Artificial General Intelligence (AGI) Participation in *Imago Dei*'.

[24] For example, Vidal, 'Anthropomorphism or Sub-anthropomorphism?' Helmreich, *Silicon Second Nature*, 87, 182, 191, 202; Geraci, 'Robots and the Sacred in Science and Science Fiction'.

[25] On the Yokohama Cemetery 'monk machine', see Colors Magazine, 'The Buddhist Monk Machine'. A wide variety of news media discussed BlessU-2, the Nissei Eco robot, and the robot *puja* in 2017. For an academic assessment of BlessU-2, see Löffler, Hurtienne, and Nord, 'Blessing Robot BlessU2', which provides a clear trajectory for the study of religion within the domain of social robotics.

the physical characteristics of religious figures ('theomorphic' robots) and collaborate with individuals in religious worship.[26]

Most of this lies in the realm of very traditional western perspectives on religion. Scholars focus upon institutional forms of religion and how advances in AI prompt theological reflection. But they leave aside a critical approach to the practice of science: the occasions when AI research becomes, itself, religious or begins competing with those traditional religious institutions. I and other scholars have argued that there is something distinctly religious in the futurist visions of AI, particularly the dreams of resurrection, perfection, and immortality that I discuss in Chapter 2 of this book.[27] Those goals make pop science AI narratives a fascinating intersection of religion, science, and technology. For obvious reasons, futuristic promises of robotic immortality challenge and compete with more traditional religious systems. By revelling in this hybridity of religion and science, however, AI futurism undermines the overly simplistic narratives used to describe the relationship between religion and science: most typically that the two must be in conflict or in harmony. As scientists do religious or quasi-religious work, articulating visions of the universe that draw upon the religious traditions of their cultures, they shape more than public opinion: they reveal the ways in which religious cultures are already important to the direction of technological progress. These fascinating intersections of religion, science, and technology can be harnessed.

The blend of religion, science, and technology in visions of AI is important to current technological development and is becoming increasingly relevant to political and public discourse about AI and robotics. As such, this book has two essential goals: first, to illuminate the passage of Apocalyptic AI into an Indian context and second, to offer a glimpse into how, given that passage, we might direct our global future through cultural collaboration. To accomplish the first task requires that we unpack the terrain of science and technology in India and the nature of

---

[26] Young, 'Reverend Robot'; Chaudhary, 'Augmented Reality, Artificial Intelligence, and the Re-Enchantment of the World'; Trovato, et al., 'Design Strategies for Representing the Divine in Robots'; Trovato, et al., 'Religion and Robots'; Trovato, et al., 'The Creation of SanTO'; Cheong, 'Religion, Robots and Rectitude'; Gould and Walters, 'Bad Buddhists, Good Robots'.

[27] See Amarasingam, 'Transcending Technology'; Noble, *The Religion of Technology*; Geraci, 'Spiritual Robots'; Geraci, 'Apocalyptic AI'; Geraci, *Apocalyptic AI*, Geraci, *Virtually Sacred*, 96–9, 170–200.

Apocalyptic AI before we put those two together. So the first half of this book makes the second half possible.

Exploring the intersections of Apocalyptic AI in American and Indian cultures requires that we move beyond the typical academic focus on Euro-American culture (or cultures) even as that cultural matrix remains vital to the narrative of the book. Despite colonial-era engagements in the intersection of traditional Indian thought and its relationship to science,[28] for decades, the academic field dedicated to the study of religion and science almost exclusively addressed the Christian West. There were certainly investigations into religion and science outside Euro-American Christianity, but these were the exceptions that proved the rule.[29] More recently, scholars have sought to improve the scope of such investigation, and much of that effort has engaged Indian religious traditions.[30] Purushottama Bilimoria and Makarand Paranjape, for example, have noted the increased attention to this area and speculated on the future of such study.[31] This rising attention on religion and science provides vital correctives to how we understand contemporary Indian life and Indian science.[32] Among these, the contribution of Banu Subramaniam deserves

[28] For example, see Raju, 'Sri Aurobindo and Krishnachandra Bhattacharya on Science and Spirituality', which discusses these two early 20th-century Indians' approach to the question.

[29] Early examples on Indian religious traditions include the work of the physicist V.V. Raman, who began discussing Vedanta and science in the 1990s and whose work essentially culminated in his book *Indic Visions*. David Gosling attempted ethnographic and sociological inquiry into religion and science in India in the 1970s, but his work was not published for an international audience until decades later as *Science and the Indian Tradition*.

[30] Preliminary efforts to move beyond the western dominated discourse include the essays in Brooke and Numbers, *Science and Religion around the World* and those in Fehige, *Science and Religion*.

[31] Bilimoria, 'A Prolegomenon for all Future Dialogues between Science and Religion or Spirituality in India', 210; Paranjape, 'Science, Spirituality and Modernity in India', 13.

[32] Not all such studies show equal depth in their engagement with daily life and practice in India, but all are worth consideration. I have methodological qualms with the approach taken by V.V. Raman (*Indic Visions in an Age of Science*), but his stalwart presence in the field of religion and science from the early 1990s gave a desperately needed push towards wider historical and theological questions in the field. David Gosling (*Science and the Indian Tradition*) offers a pioneering, if problematic, account of Indian science and religion based on ethnographic and sociological fieldwork. C. Mackenzie Brown (*Hindu Perspectives on Evolution*) and Elaine Howard Ecklund and her collaborators (Ecklund, et al., *Secularity and Science*, 145–68; Ecklund, et al., 'Religion among Scientists in International Context') provide survey data on the beliefs of Indian scientists, though the former shows far greater nuance in his understanding of Indian culture. Meera Nanda's *Prophets Facing Backwards* was an important effort to engage the political dimensions of religion and science, though her primary critique really targets postmodern constructivist philosophy of science. The finest studies of religion and science in India are unquestionably the work of Johannes Quack (*Disenchanting India*) and Renny Thomas ('Being Religious, Being Scientific'; 'Atheism and Unbelief among Indian Scientists'; 'Beyond Conflict and Complementarity'; 'Brahmins as Scientists and Science as Brahmins' Calling'). Anjali Roy's

special recognition for pointing towards the political aspects of religion–science interactions; she describes how India is 'teeming' with intersections of Hinduism and science, with many of these serving explicitly political ends.[33] More recent studies bring technology into the conversation, including a vital collection edited by Knut Jacobsen and Kristina Myrvold.[34] Importantly, the inclusion of technology promotes greater attention to lived cultural realities, better allowing academics to engage with social concerns such as caste or gender politics and the transformations that modernity brings to religion.[35]

This book approaches AI from within the academic discipline of religious studies. The discipline does not offer any particular method; it is instead more of a worldview, one that takes the phenomena of religion seriously. To exercise that, a variety of methods will be on display through the book. First, I borrow on my ethnographic fieldwork, conducted across five months in India during 2012–13, a short trip in October of 2016, and five months during 2018–19. During these periods, I lived on the campus of India's finest scientific research institution, the Indian Institute of Science (IISc). On my first visit, I was formally affiliated with IISc. On my third visit, I was affiliated with the adjacent National Institute of Advanced Studies. In conducting observations of and interviews with scientists, I follow the insights of scholars in science and technology studies, specifically that we can and ought to study the social dynamics of scientific research communities. I also engage in literary studies of science: I take both science fiction and popular science to be crucial modes through which the public thinks about technology. Both are also domains

---

essay, 'Faith Outside the Lab' is a seminal work as well. Although hers is not a 'religion and science book', Tulasi Srinivas (*The Cow in the Elevator*) offers insightful data about religion, science, and technology in contemporary India. Like these last three authors, I have attempted to engage Indian intersections of religion, science, and technology through ethnographic study (*Temples of Modernity*; 'Religious Ritual in Scientific Spaces'), including in a collaborative essay with Thomas (Thomas and Geraci, 'Religious Rites and Scientific Communities').

[33] Subramaniam, *Holy Science, passim*, direct quote of 'teeming' from 136, which describes environmental initiatives.

[34] Jacobsen and Myrvold, *Religion and Technology*.

[35] On caste, see Sangupta, 'Changing Hindutva by Technology'. On gender, see Luchesi, 'Modern Technology and Its Impact on Religious Performances in Rural Himachal Pradesh', 116–21. On transformations, see Luchesi, 'Modern Technology and Its Impact on Religious Performances in Rural Himachal Pradesh'; Jacobsen, 'Pilgrimage Rituals and Technological Change'.

employed by scientists to think about technology. Computer scientists, for example, have reflected on their own usage of science fiction in the technical field of human–robot interaction.[36] In an essay in the journal *Nature Machine Intelligence*, Stephen Cave and Kanta Dihal note the significance of fictional and non-fictional depictions of AI in science and pop culture as well as the important ways in which both fiction and non-fiction visions of AI inevitably fail to sever dreams of human salvation from disenfranchisement and damnation.[37] Given this, there must be increasing pressure on humanistic and social scientific research to engage with science fiction as an actual part of the scientific process. I identified English language pop science and science fiction, recognizing that doing so puts strenuous limits on my coverage of Indian culture. The limits imposed by my language skills notwithstanding, both pop science and science fiction provide meaningful access to the cultural reception and use of science and technology. Pop science magazines such as *Dream 2047* and *Resonance* familiarize their readers with both historical achievements and contemporary priorities in science. Science fiction interprets science and technology for public entertainment and is thus an important location for understanding people's expectations about technology. As Tarun Saint notes in the introduction to *The Gollancz Book of South Asian Science Fiction*, there can be discontinuities and disagreements—especially regarding the status of human beings and cultures—between futurism and science fiction.[38] Thus, whether we consider the U.S. or India, the published futures of science, science fiction, and pop science all concern us.

My studies of science and scientific culture indicate that there are culturally subliminal processes that contribute to scientific outcomes. The often unconscious habits of mind and action that are derived from religious culture (even by those who are not, themselves, outwardly religious) shape our priorities and our strategies. In the 21st century, we commonly hear that humanity has taken over evolution; among advocates of Apocalyptic AI, we hear that technological development is the evolutionary future of humanity and that we will soon be posthuman amalgams of machine and, if not the flesh and bones of humanity, at least

---

[36] Mubin, et al., 'Reflecting on the Presence of Science Fiction Robots in Computing Literature'.
[37] Cave and Dihal, 'Hopes and Fears for Intelligent Machines in Fiction and Reality'.
[38] Saint, 'Introduction', xxvii.

the uploaded human minds. The truth of this proposition remains un-tested. It will be clear later in this book how the origins of such claims have more to do with religious beliefs and practices than they do empir-ical demonstration. In a world where the future is unknown but our tech-nologies have so much power, and so much room to grow, it would be wise to leverage our cultural resources towards a positive future.

Nick Bostrom notes regarding AI that 'the most useful thing we can do at this stage is to boost the tiny but burgeoning field of research that focuses on the superintelligence-control problem and study questions such as how human values can be transferred to software'.[39] On the con-trary, Dylan Evans argues that such planning is based on a logical fallacy of risk/reward systems; he suggests one look instead at the material bene-fits accrued by those who suggest we fund their efforts (i.e. Bostrom).[40] Elsewhere, I have argued that cultural and even financial prestige are at stake in pop science pronouncements about apocalyptic technological transformations, but I do not entirely side with Evans on the matter of prudence.[41] Similarly, Roberto Giulioano describes the importance of seeing how the rhetoric of enchantment pervades descriptions of AI and thus deserves attention for its plausible impact on future outcomes.[42]

While I remain agnostic, perhaps even skeptical, about the likelihood of superintelligent machines, it seems relevant to plan for the possibility. Whether or not superintelligence is possible, the design of AI cannot be severed from the question of human values, and many eminent thinkers wonder how we will align the values of such machines with our own.[43] One part of this, however, is that we must first uncover the values already at play in AI design and in our cultural expectations about AI (including those of scholars such as Bostrom); then we ought to think about the values we can best leverage. This book is committed to these tasks.

---

[39] Bostrom, 'It's Still Early Days', 127.

[40] Evans, 'The Great AI Swindle'.

[41] Geraci, 'Cultural Prestige'; Geraci, *Apocalyptic AI*, 39–71. Similarly, Roff's claim in 'Artificial Intelligence' that the moral questions of AI are little if at all different from the moral questions of technology in general and that machines are by philosophical necessity incapable of moral action seems dubious in the extreme and unhelpful in the face of plausible, even if unlikely, ad-vances in machine intelligence.

[42] Giuliano, 'Echoes of Myth and Magic in the Language of Artificial Intelligence'.

[43] As an example, see Pace, 'Debate on Instrumental Convergence between LeCun, Russell, Bengio, Zador, and More'. For a sustained philosophical argument on the question of value alignment, see Wallach and Allen, *Moral Machines*.

India might be poised to make a global contribution to AI, but doing so requires further progress in how AI gets described and deployed, not just the level of funding it receives. Although largely committed to more mundane concerns such as improving education or building smart cities, the NITI Aayog publication of India's AI strategy recognizes the global conversation about AI superintelligence.[44] A momentary reference to superintelligent AI is not, however, sufficient to articulate either NITI Aayog's AIforAll goal or to engage the disparate value systems at work for those committed to superintelligent AI. Compounding this, Pankaj Sekhsaria and Naveen Thayil show how India's other prominent policy document on technological futures (*Technology Vision 2035*) shows an impoverished notion of Indian diversity and clear evidence that some groups are included in expectations while other groups, such as farmers and mothers, remain on the margins.[45] Our expectations for AI should avoid parochial or myopic visions of society, instead providing inclusive perspectives. Advancing the policy position of 'AI for Greater Good'[46] demands clarity on the values that permeate technology locally and internationally. Some of the values we use to build a cosmopolitan, global approach to AI can be adapted from both the U.S. and India, but wise adoption first requires understanding of the values already present in American and Indian approaches to technology.

Chapter 1 sets the stage by exploring how 19th- and 20th-century Indians conceived of the relationships among history, science, and politics. The chapter shows that the colonial era prompted many Indians to think in terms of historical renewal: an end to foreign rule and a renaissance of Indian (generally thought of as Hindu) wisdom. The nature of that rebirth was contested by those who disagreed over the precise contributions of India's past and the mixture of Indian and European culture. Nevertheless, there was significant intellectual agreement that such a hybrid culture would merge Indian traditions with contemporary science and technology and lead to political freedom. For many, this view of history and the future was apocalyptic in its expectations: a new world would be born and science proved central to that formation.

---

[44] NITI Aayog, 'National Strategy for Artificial Intelligence', 15.
[45] Sekhsaria and Thayyail, 'Technology Vision 2035'. On the policy document itself, see Chapter 4.
[46] NITI Aayog, 'National Strategy for Artificial Intelligence', 19.

Indians were not alone in seeing science as key to a new world: the rise of both AI and transhumanism was key to the science fiction and pop science emergence of Apocalyptic AI elsewhere in the world. In the mid-20th century, AI emerged as a science in the U.S. and became critical to American hopes for a new world. Contemporaneous to this, though distinct in its origins, transhumanist thought took root.[47] Chapter 2 describes the rise of Apocalyptic AI within the U.S. and Europe, exploring this through the contributions of four key individuals: roboticist Hans Moravec, roboticist/cyborg engineer Kevin Warwick, virtual reality designer Philip Rosedale, and inventor Ray Kurzweil. These four, named here as iron horsemen of the apocalypse, provided key scientific, technological, and pop culture interventions. In very different ways, each is crucial to the popular belief that human minds might be transferred to machines and the world undergo a transformation from one dominated by biological lifeforms to one inherited by the machines.

Chapter 3 asks what it would take to consider the transition of Apocalyptic AI into the scientific and popular milieu of 21st-century India. Using traditional Hindu expectations about cosmic realities— even though a great many Indians are not Hindu and Hinduism is far from monolithic—this chapter experiments with how Hindu religious texts conceive of political transitions and what this means for the arrival of transhumanist thinking in India. It offers the first history of futurist

---

[47] The precise lineage of transhumanist thought is yet to be fully elucidated. It can be stated with some certainty, however, that Christian visions of technological mastery provided vital beginnings. For example, Francis Bacon's *New Atlantis* was a seminal text in setting the stage for Christians believing that technology would allow them to transcend biological limits. In *The Religion of Technology*, David Noble traces a history of Christian speculation towards human and cosmic perfectability from the 12th century CE onwards. More recently, the influence of Russian Cosmism bore fruit in science and technology but also in religious and philosophical thought: the influence of the cosmists on Teilhard de Chardin (e.g., see *The Phenomenon of Man*) is easily spotted (see Young, *The Russian Cosmists*, for an introduction to Cosmist thinkers and Groys, *Russian Cosmism*, for a selection of their work and an excellent introduction to them). By the mid-20th century, science fiction and the influence of Teilhard and others produced a *zeitgeist*—though one of limited reach, perhaps—that encouraged landmark transhumanist texts such as Robert Ettinger's *The Prospect of Immortality* and *Man into Superman* and those of FM-2030 (né Fereidoun Esfandiary); for examples of the latter, see Esfandiary, *Optimum One*; Esfandiary, *Upwingers*. But it was in the late 1980s that transhumanist communities manifested on the Internet and began a process of outreach to broader pop culture. Of course, all of the philosophical and pop science contributions to transhumanism were filtered through, responses to, and often emergent from science fiction (one analysis of this dynamic can be seen in Geraci, 'There and Back Again'). More on this history should be clear in subsequent chapters.

speculation in 20th-century India and, drawing on pop science and science fiction, the shifting dynamics of such interests in the 21st century.

Having explored the historical realities and present dynamics of transhumanist thinking, especially Apocalyptic AI, in the U.S. and India, I turn to the importance of such inquiry in Chapter 4. This chapter raises questions of social control that are fundamental to the implementation of AI. The ideologies used to justify and sustain public and private use of AI matter because they incline users towards specific goals. It is important to see how the differing religious and cultural environments of the U.S. and India can offer separate and worthwhile tools in the formation of a global AI agenda. Recognizing that both American and Indian traditions can have negative or positive consequences, the chapter explores how the best of each can be brought to bear and suggests that a wider discussion of AI values could be leveraged in global policymaking.

The final chapter of the book describes how the rise and dissemination of AI, whether in Austin or Bangalore, produces a certain kind of world. The visions of technology that give strength to AI will bear partial responsibility for what kind of world emerges. So we must think with care about the technological values that are already present or emerging in our time, and we must consider how best to ensure that those values are productive ones. In the end, I believe that shared responsibility and shared values can emerge through global collaboration. To build a single vision that accommodates every human culture is inconceivable; nevertheless, to advance AI in clear awareness of the religious and cultural contexts in which we deploy it offers opportunities to use AI to advance human flourishing.

1

Waiting for the End of the World

Technology, History, and the Indian Struggle for Independence

Introduction

The end of the world heralded by artificial intelligence (AI) has a history. The next chapter briefly traces the connections between apocalyptic visions of AI and the religious visions of Judaism and Christianity. But to understand the transition of such ideas into and out of India, which is our primary interest, requires some consideration of how Indians perceive their cultural traditions, the role of science in keeping with them, and the permutations of apocalyptic visions in modern India. A look at recent history reveals the existence of political apocalypticism in 20th-century India. To drive this book's investigations—which rely upon specific modes of scientific, religious, and cultural interaction—this chapter takes an historical perspective and sets up the dynamics that allow us to understand how transhumanist visions are being adopted and transformed by Indians.

Traditional Hindu models of time and history provide key tools for considering the futuristic expectations of transhumanist promises. Specifically, by tracing the historical moves that brought industrial technology to colonial India, one notes how popular conceptions of technology never lose sight of traditional cultural expectations. Colonial-era Indians interpreted technology through a nationalist lens (one absolutely vital to discarding the colonial yoke of British control) and produced a union of religious and scientific discourse. This frames 20th-century thinking about technological progress and the social and political impact of technology. Ultimately, the integration of contemporary technology with traditional Hindu concepts of time promoted a view of historical

Futures of Artificial Intelligence. Robert M Geraci, Oxford University Press. © Oxford University Press 2022.
DOI: 10.1093/oso/9788194831679.003.0002

progress that predicted an end to political domination and the onset of a new world.

It is beyond reason to 'begin at the beginning', but a study of 21st-century science and culture ought to locate itself in the context of history. This chapter explores how visions of technology can integrate apocalyptic expectations as part of a political movement, in this case that of Indian independence. Alexander Ornella shows how a network of technological objects, a search for the sublime, a focus on narratives and aesthetics, and other features constitute a 'technological imaginary'.[1] It is something of this sort that requires description: the independence movement adopted traditional Hindu conceptions of time and took modern technology to be an indicator that the near future would bring an end of domination and the onset of a new world. Two important things emerge from this: a vision of technology at the end of the world and the political movements of liberatory futurism. Both will be vital resources for anticipating India's response to Apocalyptic AI and its contributions to 21st-century technology.

Scientific empowerment

Political and scientific independence were entwined in the colonial era, as British technical power was both a mechanism of and part of the rationale for British control. When the British arrived, there were technologically advanced sectors of the Indian economy and Indian goods were sought-after throughout the world. The value of such products and the wealth that had accumulated in India were, of course, primary drivers in the British desire for political and economic hegemony there. After a few decades in India, however, the British shifted their strategy to one of pure exploitation. 'As the East India Company consolidated its territorial control, it slowly shed its character as a body of traders whose eyes were on quick and ill-gotten profits, and settled down to fashion a despotism aimed at developing and exploiting the territory's resources efficiently and systematically.'[2] As the British consolidated their power and transformed

---

[1] Ornella, 'Towards a "Circuit of Technological Imaginaries,"' 322.
[2] Prakash, *Another Reason*, 3.

the Indian economy into a servant of imperial economic interests, India experienced a strong decline in traditional technologies, the quality of Indian goods, and the national confidence that depends upon a country's sciences and economic networks. Thanks to this process, Indians saw themselves as living in a degenerate age, a perspective that leant itself to apocalyptic imagination. Before we consider the apocalyptic worldview that emerged out of British colonialism, however, we must briefly detail the scientific and economic process that makes it attractive—the forced collapse of indigenous techniques and the pervasive experience of political and cultural alienation which drew on that collapse.

In its worst excesses, British control and the adoption of European language and social contracts established a problematic colonial consciousness—Indian philosophers and other thinkers challenged their own worldviews. Many held that English education and language compromised their own philosophical exercises; for thinkers such as K.C. Bhattacharyya (1875–1949), the adoption of English amounted to colonial subjection even as they voluntarily engaged in that process.[3] A century later, S.N. Balagangadhara describes a colonial hangover, claiming that western influences continue to prevent Indians from perceiving the world in authentic ways.[4] Recognizing this, Ashis Nandy argues that the internal psychology of submission is the most pervasive form of domination: 'particularly strong is the inner resistance to recognizing the ultimate violence which colonialism does to its victims, namely that it creates a culture in which the ruled are constantly tempted to fight their rulers within the psychological limits set by the latter.'[5] The language and education instituted by the British provided Indians with British tools and techniques that were often ill-suited to clarifying Indian life or the needs of colonized Indians.

In their excellent account of early 20th-century Indian philosophy, however, Nalini Bhushan and Jay Garfield point towards the incomplete dominance of colonial consciousness. They argue that despondent thinkers from the colonial era and contemporary India overstate the

---

[3] Bhattacharyya, 'Svaraj in Ideas'. For a comprehensive engagement with this philosophical problematic, see Bhushan and Garfield, *Minds without Fear*, 7–19.

[4] Balagangadhara, *Reconceptualizing India Studies*, 69.

[5] Nandy, *Intimate Enemies*, 3. I find Nandy's Freudian approach—and its all-encompassing psychosexual dynamics—to be incomplete, but his larger conceptualization of the colonized mindset offers a key framework for thinking through the process of orientalism by both colonizers and colonized.

power of colonization, and they show that colonial era philosophy made original contributions to global philosophy.[6] While bright spots such as creative philosophy exist, broadly speaking the adoption of British expectations and worldview happened to the detriment of local confidence and self-understanding.

Colonial power, science, and technology grew together for the European nations. British control over the subcontinent ultimately depended upon sciences such as botany, bacteriology, railroads, surveying, philology, archeology, and even comparative religion. For example, the botanical networks that enabled British production of quinine outside of Portuguese control extended the reach of their military.[7] In similar fashion, British mercantile groups extended their own power when their botanical gardens and experiments led to the production of tea in India, an achievement that allowed them to circumvent China.[8] Studies in bacteriology promoted the botanical research that advanced British resistance to disease.[9] It is obvious how railroad and telegraph networks strengthened colonial power, but it is equally the case that the British (first through the East India Company and subsequently through direct rule by the crown) used less tangible tools such as philology and comparative religion. Both produce social power, and it was in the context of justifying colonial control that comparative religion emerged as a discipline.[10] The location and study of ancient Indian manuscripts gave the British opportunities to define reality for India, ostensibly in the service of local tradition but functionally to create systems of power that the British could use to extend their control. Referring to such investigations, Gyan Prakash concludes that the British 'furnished a body of empirical knowledge with which they could represent and rule India as a distinct and unified space.'[11] Eventually, the rise of technical power through the

---

[6] Bhushan and Garfield, *Minds without Fear, passim*. The cosmopolitan world of colonial Indian philosophy could be seen as a precursor to this book's argument about culture, values, and artificial intelligence.

[7] Baber, *The Science of Empire*, 170–3.

[8] Baber, *The Science of Empire*, 169.

[9] Chakrabarti, *Bacteriology in British India*.

[10] See Chidester, *Savage Systems*, 34–52; Chidester, *Empire of Religion, passim*, esp. pp. 3–4. Similarly, archeology progressed in British India as a strategy to narrate the subcontinent's history as a unified one, which gave British colonizers the power to define the history of India and, therefore, its present; see Bhushan and Garfield, *Minds Without Fear*, 93.

[11] Prakash, *Another Reason*, 4.

industrial revolution and the rise of modern science promoted a sense of superiority among and about the British.[12] That is, the growing disparity between British technology and Indian technology led to British self-aggrandizement but also Indian feelings of inferiority, an experience to which we will return shortly.

When the British arrived in India, however, there was little justification for a racist colonial mentality. Although some scholars have assumed that Indian science atrophied under Mughal rule, there is little reason to believe this.[13] Often, such claims about scientific devolution are couched in the rhetorical divide between Hindus and Muslims. For example, Hindu kingdoms receive recognition for their contributions to science through Vedic and post-Vedic mathematics; the astronomy of Aryabhata, Bhaskara, and others; metallurgical advancements in the forging of iron and casting of sculpture; and medical innovations in surgery and Ayurveda.[14] Some of these domains saw reduced innovation in later centuries, encouraging commentators to see scientific decline.

Historical evidence does not really support the claim that science died under Muslim rule, however, as royal patronage of scholarship persisted during the Mughal era. A prime example would be the astronomer-king Jai Singh (1686–1734), who built observatories and continued astronomical investigations (though he curiously rejected heliocentrism).[15] For an example of Muslim contributions to Indian science, one might also look to the improvement of rockets under Tipu Sultan (admittedly at a time when the British were wresting control away from indigenous rulers and not during the heights of the Mughal Empire). It was, in fact, during the colonial era that India's technological strengths suffered.

Europeans did not begin the colonial era with faith in their own technological superiority. In the 17th century, European geographers actively sought to diminish the conceptual distance between Europeans and non-Europeans, collapsing differences between them; it was between 1670 and 1730 that increasing exoticization appeared in maps and geographical

---

[12] This trend applied to European colonialism generally; see Adas, *Machines as the Measure of Man*, 133–98.

[13] Baber describes a number of sciences and technologies that were maintained or expanded during Muslim rule in medieval India; *The Science of Empire*, 53–105.

[14] For an effective survey of Indian contributions to science, see Subbarayappa, *Science in India*.

[15] Baber, *The Science of Empire*, 85–91.

books, especially with an eye towards pictorializing native subservience to Europeans.[16] Europeans came to perceive themselves as technologically superior only after that became factually correct. Setting aside the advancements brought about by the industrial revolution (itself funded via wealth appropriated from India[17] and other colonized/enslaved peoples), the process of colonial domination produced scientific advancements for Europeans. Empire made certain kinds of sciences necessary, and hence possible.[18]

Burgeoning European strength was made starkly apparent by the systematic disenfranchisement of Indian capability. For example, the British dismantled Indian libraries, parceling out their contents to European collections, and ended state patronage for scientific and technical work (both Muslim and Hindu rulers considered this to be part of their princely obligations; the East India Company and subsequently the British crown saw it as opposed to their interests).[19] Educational policies compounded the problems created by dismantling the patronage system: government action effectively prohibited Indian scientific education and employment beyond that of low-level technicians.[20] In the realm of practical technologies—such as textiles, mining, rocketry, and metallurgy—state economic policies promoted British technical innovation while impeding that of India; for example, Indian exports were made nearly impossible through tariffs and even a temporary ban.[21] Commodities once valuable across the globe faltered under the legal onslaught of colonial power.

Ultimately, as the British consolidated their control over India, their respect for it lessened: India went from a place of awe to become a place of British disdain by the end of the 18th century.[22] By that time, scientific

---

[16] Schmidt, *Inventing Exoticism, passim*, especially 13–6.

[17] Nehru, *The Discovery of India*, 323.

[18] Gottschalk, *Religion, Science, and Empire*, 45 (on the example of cartography, see p. 62).

[19] Parthasarathi, *Whey Europe Grew Rich and Asia Did Not*, 253–4; see also Chakrabarti, *Western Science in Modern India*, 87; Subramanyam, *Europe's India*, 40–1; Baber, *The Science of Empire*, 93.

[20] Lourdusamy, *Science and National Consciousness in Bengal*, 27, 37, 41–2, 84.

[21] Baber, *The Science of Empire*, 117–9; Parthasarathi, *Whey Europe Grew Rich and Asia Did Not, passim*. Prior to these colonial policies, Indian goods—especially textiles—were highly sought after across the globe. Altman (*Heathen, Hindoo, Hindu*, p.4) notes their value in America, while both Adas (*Machines as the Measure of Men*, p. 41–53) and Parthasarathi (*The Transition to a Colonial Economy*, p. 5) note the value of these commodities in Europe, the Middle East, and Africa.

[22] Gottschalk, *Religion, Science, and Empire*, 139–40.

knowledge (or its absence) became a British justification for colonialism.[23] In the world of science and technology, the colonizers suffused
their rhetoric with 'essentialist attributions to Indian culture ... suggesting
that Indians were "temperamentally unfit" for the pursuit of modern
science.[24]

The elimination of Indian technical know-how and scholarly inquiry
during the colonial era was a vital component in building the colonized
mentality of 19th- and 20th-century Indians. Gandhi, for example, believed that 'we keep the English in India for our base self-interest. We like
their commerce, they please by their subtle methods, and get what they
want from us.'[25] Given that the military was largely composed of native
Indians, British rule certainly depended upon the willingness of the locals. But that willingness emerged out of more complex social dynamics
than the economic interests of a few—the interests of the wealthy cannot
explain the broader acceptance of English rule. Insofar as Indians were
made to feel inferior and grateful to the British, however, they became
complicit in their own victimization. For many in science, this led to a
double life that required both recognition from western scientific institutions and a brash rejection of it, a desire to establish a uniquely Indian
mode of science. Ashis Nandy exposes this dynamic using the life and
work of J.C. Bose, whose contributions put him at the forefront of global
science in the late 19th and early 20th centuries but whose conflicted relationship with colonial status hierarchies frustrated him throughout his
life.[26]

The British Raj employed a host of strategies that ensured the scientific subjugation of India, from suppressing scientific education to rejecting Indian applicants for good jobs. 'Of all the disabilities imposed
by the colonial situation, the most serious was the lack of adequate opportunities, both in terms of scientific education as well as employment.
The government, given its priorities, was more interested in producing
lower-grade technicians and mechanics for its many colonial projects.'[27]

[23] Baber, *The Science of Empire*, 17–8, 95, 214–5.
[24] Lourdusamy, *Science and National Consciousness in Bengal*, 15. This prejudice ran so deep
that it could apply even at low technological levels: the British considered migrant peasants incapable of the industrial routine of urban factories where they sought work (Basu, Strikes and
'Communal' Riots in Calcutta in the 1890s," 950).
[25] Gandhi, *Hind Swaraj*, 38.
[26] See Nandy, *Alternative Sciences*.
[27] Lourdusamy, *Science and National Consciousness in Bengal*, 27.

The government consistently ruled out significant advances in scientific education and practice, such as in its Universities Act of 1904, which was actively anti-science, and by its refusal to fund local scientific groups, such as Mahendra Lal Sircar's Indian Association for the Cultivation of Science.[28] This problem existed even in the social sciences: the Asiatic Society would not allow Indian members until 1829 even though its members almost exclusively learned Sanskrit and other Indian languages from Indian pandits.[29]

British interest in Hindu scriptures and traditions also pushed Indians away from science and technology. Calling themselves Orientalists, British intellectuals who were enthusiastic about India presumed it to possess great spiritual insight. William Jones, a judge in Calcutta and the founder of the Asiatic Society in Bengal, learned Sanskrit and translated the *Bhagavad Gita* and other texts for European readers. There were many outcomes from European research into Indian texts and languages, and among these was the way it promoted the worth of ancient Indian textual traditions. Rather than taking pride in now-forgotten technological and economic strengths of the 16th century, Indians could value themselves for millennia-old spiritual insights. Of course, this wasn't always a glorious thing for Indians. Knowing too little of India's history, Mahendra Lal Sircar mistakenly decided that Indians had resorted to 'empty speculations' in philosophy rather than learning science and technology.[30] In general, however, the Orientalist research agenda of translating ancient Indian manuscripts led to a valorization of Indian religion and spirituality as India's contribution to global history.

Thus, the colonial-era academic research paradigm of Orientalism became the ongoing intellectual oppression of orientalism. The first usage refers to historical individuals and their studies of Eastern religious traditions. But the political ideology integrated into that paradigm led academics, policymakers, and the general public to establish an identity between eastern cultures and 'the spiritual' while simultaneously defining western cultures

[28] Lourdusamy, *Science and National Consciousness in Bengal*, 58, 197. See also Nandy, *Alternative Sciences*, 37–8.

[29] Lourdusamy, *Science and National Consciousness in Bengal*, 43. Importantly, however, there were sociologists and anthropologists from L.K. Ananthakrishna Iyer (1861–1937) onwards who made vital contributions to the understanding of Indian communities; for an overview of such work, see Uberoi, Sundar, and Deshpande, *Anthropology in the East*.

[30] Lourdusamy, *Science and National Consciousness in Bengal*, 62.

as modern, rational, and scientific. This dynamic, most famously explored by Edward Said,[31] was a rhetorical tool that sustained and expanded colonial power. By equating Indians (as well as other peoples) with spirituality, European colonizers dismissed them as primitive and excluded them from modernity. Orientalism was thus a readymade justification for colonization.

The orientalist measure of India meant that if religion and philosophy were the land's significant contributions, it was only India's past which had meaning, not its present. In their consideration of India's role in modern life, few contemporaries disputed Macauley's oft-cited claim that 'a single shelf of a good European library was worth the whole of the native literature of India and Arabia'.[32] Nineteenth-century Indian historians willingly adopted this position, arguing that modern Indian life could be purged of irrationalities and obscurities compounded over recent centuries and return to its past, one glorified in the perspective of Indians steeped in the orientalist worldview.[33] Orientalist investigations of the past were not conducted with the expectation that Indian thought would advance European thinking; it was believed that European thinking could fix the problems Europeans ascribed to India.

Eventually, orientalism became an echo of what Marx labeled the 'sigh of the oppressed': Indians took up orientalist stereotypes to fend off oppression. Partha Chatterjee notes that Indians' weaknesses vis-à-vis the British, particularly their political impotence, forced them to seek self-worth in the ancient spiritual and philosophical traditions that Sircar decried and that the British allowed as India's only contribution to world history. Chatterjee shows how Indian nationalists looked for an inward, spiritual domain that they could control, separate from the public, political sphere that the British ruled.[34] This sovereign spiritual domain then became the root of public consciousness precisely because it denied the British colonizers complete superiority. In this way, adopting the orientalist mentality makes sense even though it obviously fails to solve the problem it addresses. Unfortunately, it also alienated many Indians from science: Nandy describes this dilemma by stating 'it was India's westernized middle classes which, having internalized the western concept of the

---

[31] See Said, *Orientalism*.
[32] Quoted in Bhushan and Garfield, *Minds without Fear*, 46.
[33] See Chatterjee, 'History and the Nationalization of Hinduism', 134–5.
[34] Chatterjee, *Nationalist Thought and the Colonial World*, 54–84.

primacy of science, felt humiliated because they felt they were inferior in the natural sciences and because they no longer believed in their own culture's hierarchy of knowledge.[35] Even as they defended the merits of their spiritual traditions, they had difficulty seeing those as meaningfully competitive with the scientific traditions from which they were excluded.

Supposedly, however, India's spiritual superiority held in potential a reversal of political fortunes and a conquest over the conquerors. For example, in *The Life and Teachings of Swami Dayananda*, Vishwa Prakash writes that 'the west may take the east as it desires, but the east is still spiritually very great. Very soon the west would be tired of its materialism, it has already grown sick and when materialism is swept away, spiritualism will have its hold'.[36] The desirability of ancient India's philosophical and religious accomplishments both presumed and indicated that 'western rationality' required a counterweight. This assumption, while denigrating Indians as non-modern and non-rational, created space for Indians to assume superiority in the domain of spirituality and thus on the world stage.

Presenting India's contribution as the counterweight to western materialism maintained the prejudicial disparagement of Indian scientific and technical thinking and relegated India's achievements to the past. In keeping with this, even efforts at recovering India's scientific achievements—such as P.C. Ray's *A History of Hindu Chemistry* and Brajendranath Seal's *Positive Sciences of the Ancient Hindus*—located those achievements squarely in the ancient world; no attention focused on India's early modern textiles or metallurgy or any of its other pre-colonization strengths. Seal, in fact, writes that ancient Hindus had a 'rigorous scientific method' that was 'genuinely and *positively* scientific' but that it was 'arrested at an early stage'.[37] That Indian arts and crafts, for example, might have depended on a rigorous method also does not appear to have occurred to Seal or many of his contemporaries.

Nevertheless, nationalists borrowed the real scientific achievements of India's ancient past to justify political and social change. Widespread interest

---

[35] Nandy, *Alternative Sciences*, 38.

[36] Prakash, *Life and Teachings of Swami Dayanand*, 300. Chatterjee discusses the attempted unification of western industry and eastern spirituality in *Nationalist Thought and the Colonial World* (p. 73).

[37] Seal, *The Positive Sciences of the Ancient Hindus*, 244, emphasis original.

in and emphasis upon scientific education began in the 19th-century and pro-nationalist thinkers became directly involved in scientific institution building. Mahendra Lal Sircar's Indian Association for the Cultivation of Science, founded in 1876, was an effort in this direction, as was the uphill struggle by J.N. Tata and others to establish the Indian Institute of Science (accomplished in 1909).[38] Similarly, the National Council on Education and the publication of science periodicals such as *The Dawn* provided scope for a resurgence in scientific activity and reflection.

For these institution builders and their contemporaries, scientific study became a moral and political imperative.[39] The late 19th and early 20th centuries saw a 'juxtaposition of three discourses: that of reformist religion; utilitarian, empirical science; and emerging nationalism. These three discourses are not mutually contradictory but interpenetrate'.[40] The transition from colonial control to independence was thus a complex process that in many regards rejected the colonial consciousness that disparaged Indian traditions and culture. The political will towards self-rule depended on Indians who gave renewed life to India's technical culture without discarding India's past. Nalini and Bhushan describe how Indian philosophy articulated a rebirth of Indian culture,[41] a process that applied equally to the role of science in the national movement.

New and old religious traditions gained strength, thanks to the orientalist movement towards inner spirituality. Reform movements such as the Brahmo Samaj and Arya Samaj were clear successes in this regard, and both took up the nationalist cause. Paradoxically, middle-class Indians who sought for themselves English education and literacy were also drawn to religious movements that rejected these values. This explains, for example, the popularity and influence of Ramakrishna Paramahansa, the Bengali saint so valued in Bengal's *bhadralok* community, especially by those individuals who met with little success in the

---

[38] On Sircar's association, see Lourdusamy, *Science and National Consciousness in Bengal*, 56–99. On the founding of the Indian Institute of Science, see Subbarayappa, *In Pursuit of Excellence*.

[39] Bassett, *The Technological Indian*, esp. 1, 7, 21, 67; Raina and Habib, 'The Moral Legitimation of Modern Science', 13–6.

[40] Habib and Raina, 'Copernicus, Colombus, Colonialism and the Role of Science in Nineteenth Century India', 64. The title, with the misspelling of Columbus, is correct.

[41] Nalini and Bhushan, *Minds without Fear*.

British administrative and economic regime.[42] The lure of Hindu saints
and religious movements was especially powerful in 1870s and 1880s as
the dream of Indian 'improvement' through British help became increasingly unlikely.[43]

In sum, the commodities and wealth of India lured European traders,
who eventually switched from acquiring luxury goods to conquering the
locals. Colonial control unilaterally supported European interests. It deci-
mated local knowledge, exacerbated the technological disparities created
by industrialization in Europe, bled wealth from India to help fund that
very industrialization, and produced in both the British and Indians a
belief that Indians had been left behind in modernity. In the minds of
both Europeans and Indians, India became synonymous with spirituality.
Searching in their own past, Indians sought to recover some of what had
been lost, whether in religious texts or scientific knowledge. The mixture
of market disruption, technological deskilling, orientalist prejudice, and
faith in an Indian renaissance may appear contradictory, but it emerged
naturally from the contradictions and discontinuities forced on India by
colonization. That mixture promotes a view of history and time through
which religious views were ingrained within politics and science.

Orientalism and the arc of history

Because so many people built Indian self-esteem on religion and spirituality, religious beliefs became politically relevant; this included traditional views of time and its end, which helped explain scientific progress
and the colonial state. The independence era drew on Hindu religious
visions of time in its formulation of history and historical changes to explain the present and predict the future. Specifically, the idea that history
is cyclical and divisible into *yugas* governed how many Indians conceived
the political domain. Occasionally, Indians looked at history as a linear
process akin to views common elsewhere in the globe;[44] but far more

---

[42] Sarkar, '"Kaliyuga," "Chakri," and "Bhakti,"' 1544. The term *bhadralok* refers to 'gentlefolk'
or 'well-mannered' people, specifically the upper-caste, professional community in 19th and
early 20th century Bengal.

[43] Sarkar, '"Kaliyuga," "Chakri," and "Bhakti,"' 1547.

[44] Although public conversations about politics, science, and religion typically utilized the
yuga cycle to describe history in 19th- and 20th-century India, there were Indians who adopted

commonly, Indians explained progress in science and politics through recourse to the cyclical *yugas*. As a result, the traditional view of cyclical time brought its eschatological vision, its interpretation of the end of time, to the popular understanding of British rule. The entire political domain was the stage for cosmic renewal, a process that would come to engage Indian perspectives on science and technology as well. As we will see in Chapter 3, this view still has some relevance today.

Views of history naturally include suspicions about the future end of historical processes. In academic terms, thinking about the end of times involves a distinction between apocalypticism and eschatology. These are related but are not synonymous terms. Eschatology is the correct term for thinking about the end of the world, though apocalypticism often has that meaning, especially in popular usage. Properly speaking, an apocalypse is a text (written or oral) in which a heavenly agent reveals the divine plan for the world. In Jewish and Christian apocalypses, which I will describe in the next chapter, the divine plan typically includes the end of history and the creation of a perfect new world. The divine plan for ending history explains why apocalypticism is so often used to simply mean the end of the world, especially in reference to nuclear cataclysm, environmental

a linear view of history instead. They borrowed linear views of history to describe a process that leads towards a better world, and they used this to motivate political and religious participation. Christian influences had a clear and decisive influence on this process, but Indian thinkers were not committed to Christian outcomes. In some cases, this allowed Indians to anticipate a future that closely approximated a linear view of future salvation in which history leads inexorably towards greater justice. The most famous of proponent of linear history was Keshub Chandra Sen (1838–1884), a member of the Brahmo Samaj and leading member of the Bengali Renaissance. Sen introduced a dispensationalist approach to history and launched a church that syncretized European and Indian religious traditions. Dispensationalist views emerged among Christians who divided the world into a series of 'dispensations' or eras in which different principles governed reality. This approach emerged notably in the teachings of Joachim of Fiore (1135–1202), who divided history into the Age of Law, the Age of Grace, and the coming Age of the Spirit which would be the salvation of humanity after the defeat of the Antichrist. Joachim's theology prefigures that of 19th- and 20th-century Christian dispensationalists (see Boyer, *When Time Shall Be No More*, 52). Nelson Darby (1800–1882) launched the movement in 19th-century England (Boyer, 'The Growth of Fundamentalist Apocalyptic in the United States', 147–57), and thus dispensationalist theology had a direct line of passage through missionaries to India. A key voice in favor of this new perspective, Sen observed 'linear and progressive self-revelations of God in history culminating in a future spiritual fulfillment of the historical process' (Brown, *Hindu Perspectives on Evolution*, 111). C. Mackenzie Brown notes how Sen adapted European religious and philosophical trends like F. Max Müller's evolutionary approach to religion—that it moves from supposedly primitive practices which deify god in nature to more advanced visions of god in history (Brown, *Hindu Perspectives on Evolution*, 112). Sen's interest in spiritual evolution aligns with the colonized mindset that focused on spirituality, but Sen's linear vision of cosmic history failed to catch the public's imagination. Lacking the scriptural authority of cyclical eschatologies, Sen's syncretic approach to Europe and India faded away.

collapse, or in popular culture (for example, disaster-themed movies). However, just as an apocalyptic text never ends its narrative with cataclysm but instead uses that as a springboard to the emergence of a divinely favoured world, apocalypticism in its truest sense is optimistic rather than pessimistic.

Apocalypticism, as a social movement, follows a fourfold structure: (1) a dualistic worldview of good struggling against evil, (2) an experience of alienation due to evil's ascendant status during the historical present, (3) a transcendent guarantee that a new world is to come, resolving the antagonism between good and evil and offering a perpetual victory for the former, and (4) the belief that humanity will be transformed—saved from evil and given purified, angelic bodies—thereby allowing the salvation of the redeemed in this new world. Apocalyptic movements are contingent upon that fourfold structure and a revelatory text.[45]

Our definition of apocalypticism derives from the experiences of ancient Jews and Christians, so it is not surprising that Indian traditions differ in important respects. Eschatological thinking does not necessarily require divine revelation or the precise structure of apocalypticism, and thus it is usually better to use eschatology rather than apocalypticism in referring to cosmic cycles of renewal in Indian traditions. Typically, Indian eschatologies interpret time as cyclical rather than linear, though definite exceptions exist. In cyclical perspectives, the end of time means a restoration to a prior (though impermanent) state of grace. Linear approaches, on the other hand, tend to envision history as a process of inevitable progress leading to a final outcome. During the British Raj, both linear and cyclical views of time (occasionally even apocalyptic thinking) helped Indians cope with the realities of colonial rule and provided hope for a new India to come.

Traditional views of history as cyclical rather than linear dominated India's high cultural landscape and perspectives on time. We have little information about what peasants and other low economic status groups may have thought about time across much of Indian history, so we are forced to reckon with calendar cycles prominent among the educated and powerful.[46] Among these latter groups, the *yuga* cycle is well attested. The

---

[45] For a detailed argument on the structure of apocalyptic thinking, see Geraci, 'Apocalyptic AI', 140–6; Geraci, *Apocalyptic AI*, 14–21; and, more briefly, the next chapter of this volume.
[46] See Sarkar, *Writing Social History*, 9.

cycle resembles some linear western models of history, such as the one proposed by Hesiod in ancient Greece. In both, life begins gloriously but declines through successive eras. But in India, the glorious days are in the future as well as the past. Hesiod's Greeks could never regain the virtues of their Golden Age, but Indians can anticipate a better world, even if cosmic time scales mean that world is unimaginably far away. Cosmic calendars divide history into four ages: *Satya Yuga* (also called *Krita Yuga*), *Treta Yuga, Dvapara Yuga,* and *Kali Yuga.*

In *Satya Yuga, dharma* is maintained, people live a long time, and they often have miraculous powers, such as communication at a distance. In successive *yugas,* people become more sinful and live shorter lives. There are clear class and gender dynamics here, as many sources obsess over feisty women and powerful (thus 'out of place') Shudras during *Kali Yuga.*[47] Most advocates of the *yuga* cycle believe the present world to be the *Kali Yuga* and thereby explain the presence of sin and suffering in our world. They anticipate the appearance of Kalki, the 10th avatar of Vishnu, who will arrive to end *Kali Yuga* and restore the world to *Satya Yuga.*

The *yuga* interpretation of history was born out of political travail, and the 19th- and 20th-century colonial usage of the *yuga* cycle is in keeping with its origins. Luis González-Reimann describes early incarnations of the *yuga* cycle emerging as direct result of Scythian rule in India (second century BCE to fourth century CE), and the cultural threat Brahminical communities saw in foreign domination and its accompanying threat to traditional social order.[48] Prior to this time, the term *yuga* had related but less specific meanings. In the most ancient Indian texts, such as the *Vedas* or *Brahmanas,* the *yuga* cycle does not refer to extraordinary cosmic timeframes involving hundreds of thousands of years: *yuga* refers to a human lifespan or an indefinitely long time in the *Vedas* and a cycle of a few years in the *Brahmanas.*[49] *Yuga* as cosmic era first appears in the *Mahabharata* and in texts dated to the same general time period (300 BCE to 300 CE), such as the *Manusmriti* and the *Yuga Purana.*[50] Such cosmic

---

[47] See Sarkar, *Writing Social History,* 26.
[48] González-Reimann, 'The Coming Golden Age', 106–7, 109.
[49] González-Reimann, *The Mahābhārata and the Yugas,* 6–7.
[50] For a summary of textual occurrences, see González-Reimann, 'The Coming Golden Age', 106–8. González-Reimann elsewhere challenges the idea that the earliest layers of the *Mahabharata* include the belief that the authorship takes place during *Kali Yuga;* see González-Reimann, *The Mahābhārata and the Yugas,* 102–6.

time periods were designed to predict a change of fortunes for the politically and religiously disenfranchised.[51]

Interestingly, the role of the *yuga* cycle in response to foreign inva-
sion is directly related to the extent to which invading rulers integrate
into Indian society; that is, the more invested a ruler is in setting down
roots, the less likely locals are to see new rulership as indicating cosmic
renewal. As González-Reimann points out, during Muslim rule, the *yuga*
cycle explained political conditions but was not used to claim that Kalki's
arrival was imminent or that *Satya Yuga* would soon arrive.[52] No doubt
this differing approach to the literature owes itself at least partially to the
fact that Muslims, unlike the British, came to India with the intention of
staying and becoming Indian.[53] While British rule was supposed to be
ongoing, the presence of any given British citizen was presumed to be
the opposite: one might extract prestige, power, or wealth from India, but
the purpose was to enjoy this in Europe. Muslim conquerors wanted to
occupy the land as adopted (and eventually native) children of it: they
became Indian. The British had no interest in becoming Indian, only in
enjoying the benefits of rule. So British rule prompted an apocalyptic
view of the *yuga* cycle—faith in the arrival of *Satya Yuga*—unseen for
many centuries.

With the British having consolidated power under the East India
Company, by the early 19th century, there was widespread employment
of the *yuga* cycle to justify political conditions. Referring to Mrityunjay
Vidyalankar's *Rājābali* (1809) as a key example, Partha Chatterjee explains that 'myth, history, and the contemporary—all become part of the
same chronological sequence: one is not distinguished from another; the
passage from one to another, consequently, is entirely unproblematical.'[54]
Within such a framework, the widespread presence of the *yuga* cycle in
contemporary accounts is hardly surprising. The British commissioned
Vidyalankar to write his history of India, which was thereby burdened
with British expectations, but Vidyalankar assumed the relevance of the

---

[51] González-Reimann, *The Mahabharata and the Yugas*, 97–8.
[52] See González-Reimann, 'The Coming Golden Age', 111–2. In fact, some traditions emerged
to blend Islamic and Hindu ideas, creating a composite savior figure; ibid., 112–3.
[53] On the difference between Islamic and British approaches, see Bhushan and Garfield, *Minds
without Fear*, 12, 58, 106, 110; see also Nehru, *The Discovery of India*, 328–9.
[54] Chatterjee, *The Nation and Its Fragments*, 80; see also 78, 86.

traditional *yuga* cycle and began his history by explaining it.[55] This per-spective was common enough: Sumit Sarkar argues that those who suf-fered in colonial India saw their present as a degenerate age. Especially common among the socioeconomically weak classes in late 19th cen-tury, Indians produced cheap tracts, plays, stories, paintings, and more, all characterized by Sarkar as 'Kaliyuga literature'.[56] Sarkar says that 'the myth of Kaliyuga vanished quickly from formal Indian historical writings or textbooks, but it continued to enjoy a vigorous if interestingly modu-lated life in other texts and contexts right down to the early twentieth cen-tury'.[57] Even more recently this trend held true. Ann Gold, for example, notes that late 20th-century farmers often perceived drought conditions as the direct result of sinfulness characteristic of *Kali Yuga*.[58]

The *yuga* cycle helped Indians understand and resist imperial domin-ation in popular movements as well as at the upper echelons of politics. Ishita Banerjee-Dube describes how in the late colonial period, British rule indicated *Kali Yuga* but that many believed such rule would end with the restoration of *Satya Yuga* in the near future.[59] Although apocalyptic sens-ibilities such as one sees in Europe and the U.S. may not be common in Indian history, the political alienation forced upon the Indian public during the colonial era certainly lead many to an apocalyptic mindset. Interest in *Kali Yuga* was not exclusively emergent out of colonialism: the 'caste and class dimensions' of Indian life, which were 'affected by colonialism but extending before and after it' clearly play a role in the application of the *yuga* cycle and the possibility of apocalyptic movements.[60] Nevertheless, Banerjee-Dube notes a surge in *Kali Yuga* preachers in the 19th century.[61]

In the 20th century, many Indians believed Gandhi to be Kalki, incarnate on Earth to restore the world to *Satya Yuga*.[62] Banerjee-Dube argues that a longstanding tradition exists in which the *yuga* cycle provides an expression

---

[55] Chatterjee, 'History and the Nationalization of Hinduism', 114; Chatterjee, *The Nation and Its Fragments*, 78; Sarkar, *Writing Social History*, 7.

[56] Sarkar, ' "Kaliyuga," "Chakri," and "Bhakti," ' 1549; Sarkar, *Writing Social History*, 192–5, 205. The essay ' "Kaliyuga," "Chakri," and "Bhakti," ' is reprinted in *Writing Social History* but as my original reading was with the stand-alone essay I have left all citations to it intact.

[57] Sarkar, *Writing Social History*, 15.

[58] Gold, 'Sin and Rain, Moral Ecology in Rural North India', 167.

[59] Banerjee-Dube, *A History of Modern India*, 401.

[60] Sarkar, *Writing Social History*, 215.

[61] Banerjee-Dube, *A History of Modern India*, 103.

[62] Banerjee-Dube, *A History of Modern India*, 290.

of suffering and of resistance to political difficulties, a position shared by Sarkar.[63] For example, a group of texts known as *malikas* were properly apocalyptic in being divinely revealed and one, the *Bhima Bhoi Malika ba Padmakalpa*, indicated that a redeemer would appear in 1941, apparently referencing Nehru and Gandhi.[64] That community clearly anticipated an imminent end to *Kali Yuga*, which belies the common claim that the end of *Kali Yuga* must always be centuries or millennia away. Political circumstances dictate how individuals interpret the *yuga* cycle, and this could produce full-blown apocalyptic thinking—noticeably prevalent among the economically disenfranchised (often feeling alienated from the land and their natal villages thanks to their integration into the urban colonial regime) and occasionally present even among those of greater wealth who were nonetheless alienated by foreign rule.[65] In apocalyptic readings of the *yuga* cycle, Indians sought some 'assurance of an eventual dissolution of sufferings and the ultimate triumph of good over evil'.[66]

Adherence to the *yuga* cycle and its traditional time frame that puts salvation thousands of years in the future may imply that Indians blithely accept the vicissitudes of fate but, in fact, for many Indians the realities of *Kali Yuga* served as launching point for political action. Bal Gangadhar Tilak, for example, repeatedly acknowledges the fact of *Kali Yuga* in the 1400 pages of his two-volume commentary on the *Bhagavad Gita*. Perhaps frustrated with his colleagues' patient parliamentary requests rather than direct nationalist action, Tilak deconstructs Vedantist claims that the Path of Renunciation trumps the Path of Action and justifies the latter as appropriate and the fulfillment of the *Gita*.[67] Tilak states that the problem of worldly labor occupied him from an early age;[68] perhaps he composed his commentary with the purpose of motiving Indians towards direct nationalist action.

Naturally, nationalism was thoroughly infused with the spirit of religion, a fact easily demonstrated in 19th century movements that either bordered on or fully crossed the line to apocalyptic thinking. Apocalyptic

[63] Banerjee-Dube, 'Reading Time', 156; Sarkar, *Writing Social History*, 191.
[64] Banerjee-Dube, 'Reading Time', 159.
[65] Sarkar, *Writing Social History*, 191, 201.
[66] Banerjee-Dube, 'Reading Time', 152.
[67] Tilak, *Śrī Bhagavadgītā-Rahasya Or Karma-Yoga-Śāstra* volumes 1 and 2.
[68] Tilak, *Śrī Bhagavadgītā-Rahasya Or Karma-Yoga-Śāstra* volume 1, xxiv.

thinking is fundamentally optimistic; it typically includes a near-term end to the suffering of the righteous. Struggling with the problems of colonization, many anticipated the arrival of Kalki and the new world. One early renouncer claimed to be Kalki even before the British crown took authority from the East India Company.[69] At the close of the 19th century, the uprising of Munda tribal groups connected Queen Victoria to Queen Mandodari, wife of Ravana in the *Ramayana*, and clearly expressed hope for the end of British rule, economic exploitation, and public anomie.[70] Birsa Munda, the group's leader, believed that fire would fall from the sky and that only his followers would survive into the new era.[71] Munda apocalypticism clearly drew upon Christian missionary activity but was just as much an indigenous Indian religious movement, borrowing from both Munda tribal beliefs and widespread Hindu beliefs (e.g. the end of *Kali Yuga*).[72] As the century turned over, a former school teacher from Bengal declared himself to be Kalki.[73]

These examples prefigured apocalyptic claims made by some of India's most influential independence leaders. Aurobindo Ghose (1872–1950), for example, sought to 'prepare a perfect humanity and help in the restoration of the Satya Yuga'.[74] Victoria Luker suggests that part of Gandhi's appeal lay in a perceived continuity between his work and tribal apocalyptic movements such as that of Birsa Munda.[75] This suggestion is perhaps strengthened by Gandhi's own rhetoric, such as when he declared that 'what I am seeking is the resurgence of *Satya Yuga* in India' in a 1919 speech.[76] As mentioned above, the author of the *Bhima Bhoi Malika ba Padmakalpa* predicted the arrival of India's redeemer in 1941, and he had Gandhi in mind for the role.[77]

---

[69] For references, see González-Reimann, 'The Coming Golden Age', 113.

[70] Luker, 'Millenarianism in India," 60, 61. Guha, *Elementary Aspects of Peasant Insurgency in Colonial India*, 294–7.

[71] Luker, 'Millenarianism in India', 59.

[72] Luker, 'Millenarianism in India', 54–6, 57–8.

[73] Sarkar, ' "Kaliyuga," "Chakri," and "Bhakti," ' 1551; Sarkar, *Writing Social History*, 193–4, 202.

[74] González-Reimann, 'The Coming Golden Age', 114.

[75] Luker, 'Millenarianism in India', 62.

[76] Quoted in González-Reimann, 'The Coming Golden Age', 114.

[77] There has been no shortage of modern Kalkis. One author suggested that Kalki had already arrived and that the world had progressed entirely to *Satya Yuga* (see Mookerjee, 'The Krta Era', 104). The author of this 'proof' gives away little in terms of his political or moral agenda, so it is unclear why he perceives the world as he does. But the timing of the essay's presentation (late 1930s) and publication in the 1940s is provocative, easily lending itself to political interpretation though the author suggests *Satya Yuga* began centuries before the Common Era. In the

Even with so much hope placed in the arrival of a redeemer, the past remained key to the future across India's nationalist impulse. In religious reform, for example, Bhushan and Garfield note the importance of a Vedic 'golden age' in the modernizing efforts of the Brahmo Samaj and Arya Samaj movements.[78] Structurally, the amalgam of religion, science, and nationalism remained committed to a conception of time aligned with traditional views of cosmic history and a psychological need to resist British colonizers' dismissal of India's worth. 'The Western-educated intelligentsia felt impelled to reinterpret classical texts and cast them in the language of the Western discourse. This produced the identification of a body of indigenous scientific traditions consistent with Western science. Nationalism arose by laying its claim on revived traditions, by appropriating classical texts and traditions of science as the heritage of the nation'.[79] Similar phenomena arise even in 21st-century science fiction, where alien powers of telepathy, for example, are but the re-establishment of Vedic sages' glory.[80]

Intellectual commitment to the *yuga* cycle led colonial Indians to reconcile the past and present by valorizing Indian history. Madhav Deshpande summarizes this, saying that 'in the Hindu conception, the past must have been more developed than the present'.[81] If one accepts the degeneration hypothesis—the claim that humanity currently lives in *Kali Yuga*, the nadir of civilization—then any accomplishment evident in the world today must be prefigured and even surpassed by some past variant. Consequently, 'a significant number of Indian elites defined "progress" as a movement towards achieving the heights of the glorious past'.[82] Ashis Nandy states that for Indians, the future is open only 'to the extent that it is a rediscovery or a renewal'.[83] Of course, this was a political strategy, a resistance that asserted value in the local community, but it was one that took advantage of pre-existing cultural trends.

20th century, a host of religious teachers, from Maharishi Mahesh Yogi to Sathya Sai Baba, have suggested that their teachings inaugurate *Satya Yuga* or that they, themselves, might be Kalki (González-Reimann, 'The Coming Golden Age', 117).

[78] Bhushan and Garfield, *Minds Without Fear*, 79.
[79] Prakash, *Another Reason*, 6–7.
[80] For example, Bardhan, 'Planet of Terror', 4.
[81] Deshpande, 'History, Change and Permanence', 11.
[82] Deshpande, 'History, Change and Permanence', 11.
[83] Nandy, *The Intimate Enemy*, 58. He provides an example of this in his analysis of Gandhi, for whom rearranging contemporary attitudes and practices would produce a 'new past' (p. 57).

Faith in the past triumphs of India remains inextricably connected to science and technology. In his brilliant engagement with modern Indian science, *Another Reason*, Gyan Prakash notes that belief in the Vedic origins of science and technology 'won widespread support among the Western-educated elite and became a key nationalist belief'.[84] In keeping with the belief that India's ancient religion was its key contribution to the world, those very Indian nationalists 'argued with remarkable ingenuity and deep cultural learning that the ancient Hindus had originated scientific knowledge, and that this justified the modern existence of Indians as a people'.[85]

Just as *dharma* declines from *Satya Yuga* to *Kali Yuga*, many Indians perceived technology to do so also. The orientalist faith in India's superior spirituality permitted advocates to restructure the history of technology in alignment with their spiritual past. They thereby rejected the colonial prejudice that science and technology were non-native to India, arguing that ancient Indians *already* possessed modern technology that had been lost and rediscovered in Europe. Reinterpreting ancient stories, such as the epics and *puranas*, 19th- and early 20th-century intellectuals claimed that flying vehicles, firearms, and other technologies existed in ancient India.[86]

---

[84] Prakash, *Another Reason*, 86.
[85] Prakash, *Another Reason*, 86. See also pp. 8–9.
[86] Late 20th- and early 21st-century Indians have also identified newer technologies with the miracles of mythological texts. There is a dark side to the politicization of science and technology: Narendra Modi and other Bharatiya Janata Party leaders mistakenly adopt the pre-independence belief that ancient Indians possessed modern technology. There are many reasons to deny that scriptural references indicate scientific and technological strengths, not least of which include the lack of a supporting technological ecosystem (i.e. all of the techniques and technologies that make advanced technology possible) and the lack of archeological findings in support of Vedic technology. A group of scientists and engineers at IISc refuted one supposed claim to ancient Indian flight in the 1970s (Mukunda, et al., 'A Critical Study of the Work "Vymanika Shastra"'); unfortunately that essay has been relegated to the dustbin of history in the face of 21st-century accolades for the imagined sciences of ancient India (see also Geraci, *Temples of Modernity*, 75–85). Creating a public furor (in which a great many Indians supported the prime minister), Modi declared at a hospital inauguration, 'it is said in the *Mahabharata* that Karna was not born from his mother's womb. This means in the times in which the epic was written genetic science was very much present. We all worship Lord Ganesha; for sure there must have been some plastic surgeon at that time, to fit an elephant's head on a body of a human being' (Quoted in Thapar, 'The Two Faces of Mr. Modi;' for an analysis of the public response to Modi's claim, see Geraci, 'Saffron Glasses'). One of Modi's allies, Dina Nath Bhatra, composed the textbook *Tejomay Bharat* ('Shining India') which declares that

> America wants to take the credit for invention of stem cell research, but the truth is that India's Dr Balkrishna Ganpat Matapurkar has already got a patent for regenerating body parts.... You would be surprised to know that this research is not new and

Dayananda Saraswati (1824–1883), founder of the Arya Samaj reform movement, was among the first religious reformers and nationalists
to argue in favour of India's technological priority. In *Satyarth Prakash*
(*'The Light of Truth'*), he writes: 'it is a fact, that all the sciences and arts
and religions, that are now found in the whole world, took their original
start from Aryavarta', meaning North India.[87] As examples, he claims
that ancient Indians possessed steamships, modern weaponry, and flying
machines.[88] This view of history and technology was inseparable from
Saraswati's nationalism and resistance to colonialism. Referencing a
flying machine and fan that never ceased operating—both described in
the 11th century CE text *Bhoja Prabandha*—he writes that 'had these two
inventions come down to these days, the Europeans would not have been
so puffed up with pride, as they are now-a-days'.[89]

While technology was crucial to all 19th-century interpretations of the
historical present, accepting the *yuga* cycle did not necessitate Indians believe that they lived in a degenerate age: some Indians interpreted scriptures to indicate that they were already in a period of social, political, and
scientific progress. They did not believe that humanity remained in *Kali
Yuga*, instead arguing that *Kali Yuga* had ended centuries ago and that
humanity was in the process of improvement. For them, the *yuga* cycle
was still crucial to reading history and predicting the future, but their optimism could be grounded in political and scientific processes already

that Dr Matapurkar was inspired by the Mahabharata. Kunti had a bright son like the
sun itself. When Gandhari, who had not been able to conceive for two years, learnt
of this, she underwent an abortion. From her womb a huge mass of flesh came out.
(Rishi) Dwaipayan Vyas was called. He observed this hard mass of flesh and then he
preserved it in a cold tank with specific medicines. He then divided the mass of flesh
into 100 parts and kept them separately in 100 tanks full of ghee for two years. After
two years, 100 Kauravas were born of it. On reading this, he (Matapurkar) realised
that stem cell was not his invention. This was found in India thousands of years ago.
(Bhatra, *Tejomay Bharat*, 92–3, quoted in Dalal, "Gujarat Model of Using Epics as
Facts in Education")

The reference to America and a host of uncited speculations reveals how the obfuscation of myth
and history continues to serve political ends as India seeks freedom from western domination,
this time economic and scientific rather than direct political rule. Such is true for Modi's claims
and a host of similar examples: these technological enchantments represent India's continued
search for independence, its continued faith in a world to come (for more examples of contemporary faith in Vedic technology, see Geraci, *Temples of Modernity*, 42–5, 75–85).

[87] Saraswati, *Satyarth Prakash*, 293.
[88] Saraswati, *Satyarth Prakash*, 282, 292, 312.
[89] Saraswati, *Satyarth Prakash*, 312.

underway rather than in the hope of a radical break at the arrival of Kalki. While some religious leaders implied or claimed they were Kalki come to inaugurate *Satya Yuga*, other 19th- and 20th-century thinkers simply rejected the common view that the present remained *Kali Yuga*. For these, *Kali Yuga* ended centuries prior: technological and political progress is both the result and demonstrable proof of improved historical conditions and the shift out of *Kali Yuga*.

Sri Yukteswar Giri (1855–1936) was key to the emergence of faith in historical progress during the nationalist era and post-independence period. Sri Yukteswar, perhaps best known as the guru of Yogananda Paramahansa, offered a critical reinterpretation of the *yuga* cycle that was both in keeping with the anticolonial struggle and revolutionary in its approach to history, science, and technology. Sri Yukteswar read the *Manusmriti* as indicating that the *yuga* cycle reverses itself rather than repeating itself.[90] In this theory of the *yugas*, instead of cycling from *Satya Yuga* to *Kali Yuga* and then restarting at *Satya Yuga*, the eras reverse at the endpoints: *Satya Yuga* to *Treta Yuga* to *Dvapara Yuga* to *Kali Yuga* to *Dvapara Yuga* to *Treta Yuga* to *Satya Yuga* to *Treta Yuga*, and so on, with a kind of transitional overlap between each era. This perspective has been so influential that an image search on Google for '*yuga* cycle' provides a panoply of results depicting Sri Yukteswar's version. According to his calculations, the world began processing towards *Dvapara Yuga* in 1600 CE and the onset of *Dvapara Yuga* took place in the year 1700.[91] Importantly, Sri Yukteswar uses the history of science to make his case: it is the discovery of magnetism and the laws of planetary motion, the invention of the microscope, and contemporaneous scientific and technological innovations that provide his initial evidence that historical degeneration already reversed. For Sri Yukteswar, technological limits indicate the fallen state of humanity; transcending those limits reveals progress towards a new world.[92]

---

[90] Giri, *The Holy Science*, xvii.

[91] Giri, *The Holy Science*, xvii.

[92] Of course, spiritual transcendence supersedes and encompasses scientific transcendence for Sri Yukteswar. His student, Yogananda Paramahansa implies this when he describes Sri Yukteswar as startling scientists, philosophers, and others by possessing 'precise insight into their specialized fields of knowledge;' see Paramahansa, *Autobiography of a Yogi*, 127.

Overall, nationalism incorporated technological progress and religious eschatology. A political will for independence arose requiring a shift in the world age, ending the *Kali Yuga* that made British rule possible. This depended upon a technological imaginary where technical progress is a return to past glories, a key aspect of nationalist development, and evidence for the coming of a new age, one where India sheds British control. In both politics and science, the religious view of *yugas* offered attractive explanations to colonial Indians. The mixture of religion, politics, and science was simultaneously a reflection of colonial domination and resistance to it. Both European orientalism and the Indian adoption of it distanced 19th- and 20th-century Indians from science. Whether descending or ascending, the *yuga* cycle offered an opportunity to frame Indian progress as a rebirth of past glory while making it clear that science and technology were as Indian as they were European. The *yuga* cycle provided colonized Indians with a worldview that united their desires for political and scientific independence.[93]

Conclusion

Colonization created a confusing dynamic for science and technology in India, one that provides several possible directions for the integration of AI and futurist speculation in 21st-century India. By dismantling India's craft and technology at the very moment when industrialization became possible in Europe, the colonizers forcibly relegated India to a backward status in modernity, one that Indians adopted and tried to own as a counternarrative. Through the honorifics bestowed by the British Orientalists, they created their own orientalist strategy of glorifying

---

[93] Although India became independent shortly after World War II, eschatological faith in the near future remains tied to visions of technology that depend on the *yuga* cycle and its vision of time, especially as articulated in the colonial era. It is worth noting that India's government early in the 21st century, ruled by the Bharatiya Janata Party, accepts two key propositions: that historical progress, not further degeneration, is underway and that ancient Indians possessed advanced technology characteristic of the past century. Of course, that government, suggesting a new renaissance, would gain little political purchase from believing that the world is in a state of inevitable decline and thus has welcomed the idea that the world is progressing and is thus implicitly experiencing the ascending *yugas* articulated by Sri Yukteswar. The government of India makes no specific claims about *yugas*, of course, but the 2015 award of *Padma Bhushan* to David Frawley, an advocate of Sri Yukteswar's ascending *yugas*, lends official authority to that perspective (for more on this, see Geraci, 'A Tale of Two Futures;' Geraci, 'Saffron Glasses').

India's spirituality, almost exclusively understood as long-distant historical accomplishments. Even modern technology was enmeshed in this orientalist favour of ancient religion.

British colonizers built their empire on the resources they wrested from the locals they eventually conquered outright, and this had a particular impact on Indian science and technology. The early arrivals from Europe were far more likely to base their cultural judgements on religion (seeing their own as superior) than on technology. But the growing technological might of Europe allowed the British to assert their scientific superiority and to ground their self-esteem in this. To some extent, that opened the door for European admiration of Indian philosophical and spiritual traditions. But it also promoted the idea that Indians were spiritual and mystical in a way that Europeans were not, and commensurately that Europeans were rational and scientific in ways that Indians were not. Although Indian craft and industry had been far superior to many European counterparts, by the late 18th century, these traditions had been disrupted by British economic and social policies. Even Indians lost the memory of their own previous strengths and were left to trumpet only their prowess in religious domains and long-distant history. At best, they hoped the British would assist them towards a scientific modernity.

The colonized mindset, the commitment to the past, and the dismissal of the Indian present capitalized on the view of time made popular through the *yuga* cycle. A cosmic calendar that itself sought to revitalize a conquered people became the explanation for how Indians were conquered in the present. But in their recourse to *Kali Yuga* as the explanatory principle for their suffering, Indians returned the *yuga* cycle to its original usage: an apocalyptic prediction of better days to come. But unlike 2000 years ago, this time science and technology were integral parts of the cosmic predictions.

The apocalyptic view of history combined with the urge built in by the disruption in Indian technology and culture to create a redemptive view of technology. While these specific dynamics are unique to Indian history, we see similar perspectives on technology elsewhere. In the 19th and early 20th centuries, Americans lauded the power of technology to reshape the landscape and establish an earthly Eden. They saw the growth of technological control as commensurate with a divine plan for their nation.[94] As the 20th century unfolded, much of the outward Christianity disappeared from this view, but the commitment to redemptive technology persisted through the development of entirely new technologies. Muddled by Cold War fears of nuclear annihilation and committed to Christian visions of cosmic transformation, Americans saw science as the door to a new world. The next chapter thus traces the 20th- and 21st-century developments that integrated AI with the apocalyptic commitments of American culture.

The apocalyptic vision of AI that emerged in the U.S. offers multiple competing but intertwined interpretations of history and its direction. Those interpretations and their engagement with Indian culture will occupy the remaining chapters of this book. In order to understand how a global vision of AI, one committed to human flourishing, can be constructed out of varying cultural tools, subsequent chapters describe the rise and significance of Apocalyptic AI, its initial engagements in India, and alternate narratives that Indian culture makes possible. As we look towards the possible end of the world as we know it, staring down the barrel of climate change, but also wondering what changes digital technology will force upon us, we see that global problems require global solutions. By investigating the cultural foundations for our technological imaginaries, we launch ourselves towards a better conversation about technological development.

---

[94] Nye, *America as Second Creation.*

2

The Iron Horsemen and the End of Times

A Futurist Blend of Religion, Technology, and Cosmic Transformation

Introduction

The last chapter proposes that the colonial history of India established an intersection of religion and technology that operates on Indian political life. This intersection is also the location for new cultural ideas leading Indians to consider their nation's future in world affairs and the development of technology. Specifically, the global deployment of artificial intelligence (AI) and the international transmission of ideas *about* AI affect Indian understandings of history and modernity. Such ideas provoke a response from India, so we must understand their nature, origins, and evolution. This chapter describes how contemporary approaches to AI are inextricable from goals that emerged from Euro-American religious traditions. But to understand how those views matter in India, we must first focus on the ideas themselves and consider how they arose—primarily in the U.S.—and what promises currently govern global interpretations of AI.

Descriptions of AI's future frequently revolve around human transcendence and the apocalyptic transformation of the world from one dominated by biology to one dominated by machines. These dreams percolate through science fiction, popular science, and industry, having transitioned from fringe counterculture to the mainstream in 21st-century life. Pop science and pop culture narratives about AI have considerable importance; as Roberto Giuliano notes, 'fables, fairytales, myths and science fiction stories or novels function as a higher order language. If words aid us in crystallizing phenomena, carving up perceptible portions of the

Futures of Artificial Intelligence. Robert M Geraci, Oxford University Press. © Oxford University Press 2022.
DOI: 10.1093/oso/9788194831679.003.0003

world and making it possible to communally transmit and share information about them, then art forms expand these powers of communication to even greater heights'.[1] In the case of AI, the narratives that describe our world are not just myths and science fiction; the artistic imagination is bolstered by scientific claims and even by technological designs such as online virtual worlds. The iron horsemen of the apocalypse—technologists who actively contribute to the public and scientific expectations of radical transcendence (regardless of the extent to which they accept or promote it)—include Hans Moravec, Ray Kurzweil, Kevin Warwick, and Philip Rosedale. Their work establishes a technological enchantment that has implications for the future of Indian science and technology also. The worldview they help build—what I've called Apocalyptic AI—can be limited by competition with other mythical narratives, but economic and social changes are making it possible for this blend of religion and technology to garner an increasing share of global culture.

Although it is often purported that science is value-free and that a strong demarcation separates religious and scientific practice,[2] reality is far more complex. Throughout the world, scientific practices are intertwined with a host of religious practices.[3] In the last chapter, I described the emergence of vibrant intersections of religion and technology in India; this chapter lays out the religious impulses buried in western readings of AI and robotics. A religious current flows through formal and informal discussions of robotics and AI, one that draws its strength from apocalyptic perspectives inherited from Christian theology.

Apocalyptic AI is a movement whose chief proponents are elite scientists and engineers; they argue that we live in a world where intellect and bodies are in conflict, but the present supremacy of biological embodiment (and its attendant limits) will give way to the transcendence of mind. They believe that superhuman machines will replace human beings

[1] Giuliano, 'Echoes of Myth and Magic in the Language of Artificial Intelligence', 11.
[2] The most common example of what Barbour (*Religion and Science*, 84–89) calls 'independence' between religion and science is Stephen Jay Gould's theory of non-overlapping magisteria, which Gould describes in *Rocks of Ages*.
[3] For a particularly noteworthy example from India, consider the festival of *Ayudha Puja* (generally translated as 'worship of the machines'); Geraci, *Temples of Modernity*, 112–23; Geraci, 'Religious Ritual in Scientific Spaces;' Thomas and Geraci, 'Religious Rites and Scientific Communities'. Festivals to the god Vishwakarma are clearly related to the *Ayudha Puja* festival; for analyses of these, see Narayan and George, 'Tools and World-Making in the Worship of Vishwakarma' and Narayan and George, 'Vishwakarma'.

in a radical transformation of life, intelligence, and physical matter. As part of this cosmic transition, the iron horsemen predict that human beings will merge with machines, becoming cyborgs, robots, or software avatars that occupy the radical new world to come.

Religious remnants

Scholarship in the history and sociology of science shows a wide variety of political and personal commitments that affect scientific practice.[4] Naturally, some ways of thinking that affect the development of science and technology evolved out of religious practices and beliefs. European progress in science and technology took place in a Christian context that leant political, moral, and teleological significance to them. Twenty-first-century scientists rarely speak of their work in terms of divine providence, but subliminal Christian commitments remain in cultural attitudes about science and its goals. There may be ways in which such ties inhibit science or produce unfortunate public outcomes, but this is not necessarily the case. For the present, it is sufficient to recognize the historical arc by which Christian theology was instrumental to the construction of modern science and technology.

In his well-regarded history of religion and technology, David Noble shows that Christian theology was a primary driver in the establishment of European science. In particular, a theological push towards salvation was so firmly entrenched in science and technology that even after those domains became secular they retained their theological goal of perfecting the world and humanity.[5] Noble traces the medieval desire to see technology as part of a divine plan to overcome the fall of Adam, to wage war against the Antichrist, and to bring about a perfect heavenly kingdom.[6] This was not an innocent project, as Noble notes: all of these theological matters intersected with a political will to conquer Islamic and American

---

[4] For classic examples, see Greenberg, *The Politics of Pure Science*; Harding, *Is Science Multicultural?*; and Haraway, *Modest_Witness*. For examples from India, see Nandy, *Alternative Sciences*; Thomas, 'Brahmins as Scientists and Science as Brahmins' Calling'.
[5] Noble, *The Religion of Technology*.
[6] Ibid.; see also Schaffer, 'The Devices of Iconoclasm', 503.

lands.[7] Natural philosophers of the medieval and early modern periods explicitly connected their theological goals to progress in the mechanical arts (i.e. science and engineering). While explicit Christianity drops out of the story by the late 20th century, the same language of human salvation and a transcendent kingdom to come continues to frame advanced technologies such as AI, genetic engineering, and spaceflight.[8]

The time period in which Christian salvation became submerged in a secular language of scientific progress is simultaneously that of the rising political influence of the U.S., which combined Christian theology and technical progress into its national mythos. In the American colonies and subsequently the U.S., an ideology of divine progress and technical mastery over the land constituted essential components of continental expansion.[9] This provided the ideological backdrop for continuing colonization by the Euro-American settlers. Echoing the European tradition that traces back to medieval Christian opposition to Islam,[10] the settlers believed themselves to have a divine mandate, decimating indigenous populations and seeing technological progress as a sign of heavenly favour.

By the 20th century, explicit mention of Christianity within mainstream scientific discourse had largely come to an end. It was this fact which prompted the vigorous anti-evolution movement in the U.S. For example, in their Creationist classic, *The Genesis Flood*, Morris and Whitcomb decry the godlessness of modern geology and bemoan the fact that trained geologists no longer have any need for Biblical explanations.[11] Despite the effort of Creationists and others, Christian themes of transcendence have outgrown their religious origins, rooting themselves in a public and scientific discourse that no longer requires adherence to Christianity or its god.

---

[7] Noble, *The Religion of Technology*, 24–31. I am grateful to Syed Mustafa Ali for pointing out to me that I have too often omitted the political and race conflicts built into this theological agenda.

[8] Noble, *The Religion of Technology*; see also Midgley, *Science as Salvation*.

[9] See Nye, *America as Second Creation*.

[10] See Ali, '"White Crisis" and/as "Existential Risk."'

[11] Morris and Whitcomb, *The Genesis Flood*, 116–7. Similar complaints are, of course, visible in later Intelligent Design authors who sought to obfuscate their specifically Christian agenda while nevertheless pressing for it: for example, see Johnson, 'Evolution as Dogma;' Behe, 'Molecular Machines'.

In particular, the idea that cosmic transformation would overcome the fallen state of humanity became a scientific staple in the 20th century. Across Euro-American cultures, there were interesting landmarks in this process. Late in the 19th century, Nikolai Fedorov, a Russian librarian, took the Christian desire for salvation and placed responsibility for it securely in the technological creativity of humanity. Fedorov believed that it was humanity's obligation to develop technologies for securing immortality and resurrecting the dead.[12] That is, it was no longer the obligation of Jesus to return and save humankind but rather the obligation of humanity to produce the heavenly kingdom.[13] For Fedorov, this was sanctified by his Russian Orthodox faith, but many of his followers departed from his explicitly Christian framework.[14] Transhumanist dreams remain in 21st-century Russia, and this is often in conflict or competition with Russian Orthodoxy.[15] Some of Fedorov's followers carried his technological faith to Western Europe, where it appears to have had impact on the evolutionary thinking of the Jesuit priest Teilhard de Chardin and on Julian Huxley, one of the founding figures in the transhumanist movement.

Teilhard de Chardin was resolutely, if controversially, Catholic (and his thinking remains a possible bridge between atheist transhumanists and 21st-century Christians[16]), but Huxley transformed the Christian reading on technology into a religion of its own. Huxley was the first to advocate for the word transhumanist as a description of those who sought to transcend human limits through technology,[17] and he was a founding figure

---

[12] See Young, *The Russian Cosmists*, 80–2; Bernstein, 'Freeze, Die, Come to Life', 772–5.

[13] Although I use the term 'humankind', there were a host of gendered expectations in Fedorov's account, which articulates a resurrection of 'the fathers' and their relationships with 'sons;' see Young, *The Russian Cosmists*, 48–9, 62, 87–9.

[14] On Fedorov's Christian commitments, see Burdett, *Contextualizing a Christian Perspective on Transcendence and Human Enhancement*, 25–8; Burdett, *Eschatology and the Technological Future*, 18–24. On some of his followers' disparate views, see Young, *The Russian Cosmists*, 145–76. In contemporary Russia, the transhumanist community is split between Fedorovians, who remain committed to Russian Orthodoxy and immortalists, who reject Christianity; see Bernstein, *The Future of Immortality*, 17–8.

[15] Bernstein, 'Freeze, Die, Come to Life', 766, 768, 769, 776.

[16] Steinhart, 'Teilhard de Chardin and Transhumanism;' see also Delio, 'Religion and Posthuman Life;' Burdett, 'Contextualizing a Christian Perspective on Transcendence and Human Enhancement', 29–32; Burdett, *Eschatology and the Technological Future*, 113–40. A number of important tech visionaries also point to Teilhard de Chardin as relevant and inspirational; see Au, *The Making of Second Life*, 7.

[17] Huxley, *New Bottles for New Wine*, 17; in an important historical addendum, both Cole-Turner ('Going Beyond the Human') and Vita-More (*Transhumanism*, 10) point out that Dante made a similar usage of the Latin *transhumanar*. Vita-More has noted that Fereidoun Esfandiary,

in the effort to make a religion out of scientific transcendence. 'What the world needs is an essentially religious idea-system, unity ... charged with the total dynamic of knowledge old and new, objective and subjective, of experience scientific and spiritual.'[18] Huxley argued that 'evolutionary humanism' or 'transhumanism' could be that system.

Many transhumanists, however, chose to obfuscate the religious elements of scientific transcendence, preferring instead a sterilized, anti-religious version. Drawing on the work of Huxley's friend J.B.S. Haldane, as well as mid-century science fiction authors, late 20th-century transhumanists such as Max More, Natasha Vita-More, and Nick Bostrom articulated and modernized this position.[19] For these, transhumanism is a philosophical quest to transcend our mortal limits through technology. Ultimately, however, the religious history and implications for such a search should not be forgotten.[20] Indeed, Abou Farman points out that transhumanists, butting up against the boundaries of the secular, turn to the language of spirituality precisely because transhumanism 'is the cultural zone where secular materialism overspills its bounds.'[21]

While transhumanist thinking grew slowly in science and in pop culture,[22] it made significant inroads into science fiction; these would later produce new movements such as apocalyptic perspectives of AI. Three particularly notable science fiction authors who had special influence on the nexus of transhumanist thought and AI were Isaac Asimov, Robert Heinlein, and Arthur C. Clarke. Asimov is known, of course, for his interest in robotics and human–robot interaction, and his influence on the scientific study of robotics is well documented.[23] Heinlein's Lazarus

---

a leading proponent of 20th-century transhumanism, was surprised to hear there had been prior usage of the term (see her online comment on Pablo, 'Transhumanism and FM Esfandiary').

[18] Huxley, *Religion without Revelation*, unpaginated preface.
[19] Haldane, *Daedalus*; More, 'Principles of Extropy'; More, *Philosophy of Transhumanism*; Vita-More, *Transhumanism*; Bostrom, *In Defense of Posthuman Dignity*.
[20] For a history of this transition, see Geraci, 'There and Back Again'.
[21] Farman, 'Mind out of Place', 59.
[22] Historically important mid-20th-century pop science advocates of transhumanist themes include Robert Ettinger and Fereidoun Esfandiary (discussed later in this chapter). Aside from Haldane and Huxley, scientific contributions to the transhumanist movements were less common but can be seen in remarks by some scientists, see Ulam, 'Tribute to John von Neumann;' Bush, 'As We May Think;' Good, 'Speculations Concerning the First Ultra Intelligent Machine;' and Martin, 'Brief Proposal on Immortality'.
[23] See Asimov, *I, Robot*; *The Caves of Steel*; *The Naked Sun*; *The Robots of Dawn*. On his influence, see Geraci, *Apocalyptic AI*, 52, 54.

Long became the hero of generations who sought life extension technolo-gies.[24] And Clarke was the chief science fiction advocate for two major transhumanist desiderata: using technology to evolve beyond human limits and uploading minds into machines.[25] Science fiction remained on the literary fringe until the 1990s and early 2000s, but these and other authors popularized a futuristic confluence of technological advance-ments to produce a new world occupied by new species of intelligent life. They nourished a growing faith in science and technology as rational and plausible avenues towards human salvation.

Apocalyptic AI draws heavily from the science fiction authors that promote our escape from earthly, biological matter. In both *The City and the Stars* and *2001: A Space Odyssey*, Clarke suggests that technology will enable life to escape the 'tyranny of matter'.[26] Decades later, William Gibson writes that in cyberspace, a hacker can escape 'the prison of his own flesh' in favour of the 'bodiless exultation of cyberspace'.[27] Gibson's work prompted increasing devotion to cyberspace, leading to Neal Stephenson's lavishly described Metaverse in *Snow Crash*. Here too, life in virtual reality trumps that in the physical world. Stephenson writes of his hero: 'when you live in a shithole, there's always the Metaverse, and in the Metaverse, Hiro Protagonist is a warrior prince'.[28] As one might expect, *Snow Crash* directly inspired virtual world designers, including Philip Rosedale (to whom we will shortly return).[29]

The evangelism of science fiction and pop science authors inspired the modern movement of transhumanism. As noted earlier, contem-porary transhumanism—the belief that humans can use technology to transcend our biological limits—draws on historical traditions that date back at least into the early modern period. For example, Francis Bacon fantasized about Christians extending their lifespans and mastering all

[24] Heinlein, *Methuselah's Children*; *Time Enough for Love*. Alexander notes the significance of Heinlein to transhumanists in *Rapture* (pp. 48–50).
[25] Clarke, *Childhood's End*; *2001*; *The City and the Stars*. One should not lose sight of the fact that Vernor Vinge was the first to suggest voluntary mind uploading into cyberspace rather than some other body or energy matrix (see *True Names*). Frederick Pohl was actually probably the first science fiction author to describe minds uploaded into a virtual environment in 'The Tunnel Under the World', but Pohl's description is of involuntary upload as part of a corporate dystopia. In Vinge's book, a character voluntarily chooses mind uploading.
[26] Clarke, *The City and the Stars*, 263; *2001*, 185.
[27] Gibson, *Neuromancer*, 6.
[28] Stephenson, *Snow Crash*, 63.
[29] Geraci, *Virtually Sacred*, 171–2.

aspects of nature through science and technology.[30] In the middle of the 20th century, in a crucial development for the rise of an explicitly transhumanist movement, Robert Ettinger described cryonics—the freezing of the deceased's body or head in hopes of future resurrection—as a step in the direction of immortality and then swiftly expanded to a most more robust vision of human enhancement and our evolution into a new (though still biological) species.[31] He was more than just a 'thinker in the clouds': he established the Cryonics Institute to realize his vision. Fereidoun Esfandiary, who would later change his legal name to FM-2030, produced the most wide-ranging transhumanist texts of the time and also established a group of like-minded thinkers around him in Los Angeles. Like Ettinger, he became something of a cult figure, known in certain circles through his published books and essays.[32] His cohort included Max More (né Max O'Connor) and Natasha Vita-More (neé Nancie Clark), both of whom would have an enormous impact on the transhumanist movement. With the coming of the Internet, More's Extropy Institute reached a wider audience and became crucial to sharing transhumanist ideas. Later, he became CEO of the Alcor Life Extension Foundation, which operates a cryonics facility in Arizona. Drawing on the Extropy Institute's work, the World Transhumanist Association formed in the 1990s, later becoming Humanity+. Vita-More lead the group in the 21st century as her own influence spread.

Academics in religious studies and theology have inquired into the relationship between transhumanism and traditional religions and also into the religious nature of transhumanism itself. Some of these, such as Hava Tirosh-Samuelson, find transhumanism a threat to human dignity.[33] Others find some theological room for maneuver, believing that

---

[30] See Bacon, *New Atlantis*.

[31] Ettinger, *The Prospect of Immortality*; *Man into Superman*.

[32] For example, Esfandiary, *Optimism One*; Esfandiary, *Upwingers*; 'Up-Wing Priorities'. Some of his public writings, which helped locate him as a public intellectual, engaged issues outside transhumanism: for example, Esfandiary, 'Gaining Perspective;' Esfandiary, 'The Mystical West Puzzles the Practical East'. It is quite possible that many of Esfandiary's ideas were influenced indirectly by Julian Huxley: Abraham Maslow, whose hierarchy of needs was well known, was himself influenced by Huxley (see Deese, *We Are Amphibians*, 179–80). Certainly, Esfandiary's contemporary, Ettinger, was aware of Maslow (*Man into Superman*, 124–6). It is thus at least minimally plausible that some lines of transmission to Esfandiary were similarly extant. Without attempting to demonstrate dependence, Schussler provides a comparative analysis of Maslow's psychology and transhumanism in 'Transhumanism as a New Techno-Religion' (97–9).

[33] Tirosh-Samuelson, 'Transhumanism as A Secular Faith;' 'Wrestling with Transhumanism;' 'In Pursuit of Perfection'.

some of transhumanism's essential motivations are compatible with traditional theology.[34] A wider approach to the question of religion and transhumanism appears in volumes edited by Calvin Mercer and Tracy Trothen and in one edited by Mercer and Derek Maher.[35] Organizational work by Mercer, Trothen, and others led to the creation and vigorous activity of a Religion and Transhumanism group at the annual meetings for the American Academy of Religion. The extensive academic work engaging religion and transhumanism—which far exceeds a brief paragraph's ability to encompass—reveals that transhumanist advocacy has succeeded in placing the movement into the mainstream. Indeed, Tirosh-Samuelson notes with concern that the increasing public acceptance of transhumanist thinking has taken place precisely at a time when quite a few of the academic studies of the movement critique it.[36] Other scholars are more sanguine, but the increasing prevalence of transhumanist thinking in entertainment, science, and elsewhere in public life shows that transhumanist thinking is of the age rather than that it is necessarily good or bad.

Within the transhumanist profile, a powerfully recurring theme is the one I've labeled Apocalyptic AI.[37] In short, it is the belief that future progress in AI and related technologies will bring a transformative new world of godlike machine intelligence. The iron horsemen described later in this chapter are among those central to the rise and continuation of the Apocalyptic AI movement. Just as we have seen that Christian theology was part and parcel with the development of transhumanist thinking in general, Christian—and as a precursor Jewish—traditions form the core intellectual paradigm for Apocalyptic AI. The apocalyptic traditions of ancient Judaism and Christianity provide the framework by which these advocates of robotics and AI think through our future.

---

[34] For examples, see Green, 'The Technology of Holiness'; Garner, 'The Hopeful Cyborg'; Thweatt-Bates, *Cyborgs Selves*; Fisher, 'More Human than the Human?'.

[35] Mercer and Trothen, *Religion and Transhumanism*; Trothen and Mercer, *Religion and Human Enhancement*; Mercer and Maher, *Transhumanism and the Body*.

[36] Tirosh-Samuelson, 'In Pursuit of Perfection', 204.

[37] I first used this phrase in 'Spiritual Robots' in 2006 and provided more detailed arguments in the essay 'Apocalyptic AI' in 2008 and the book *Apocalyptic AI* in 2010.

Apocalyptic origins

The salvation promised by science and technology has its roots in Christian theology, and an important part of that theological under-pinning is its apocalyptic vision of cosmic transformation. Apocalyptic thinking permeated medieval Europe, especially as Europeans engaged with Muslims in Europe, North Africa, and the Middle East.[38] Much later, American conflict with the Soviet Union produced anxieties that appear in pop culture, science, and religion.[39] Religious anxieties were distinctly different from those in pop culture and science, however, because po-tential world-ending conflict was not just mitigated but necessitated and justified by the coming Kingdom of God. Based on traditional apoca-lyptic thinking drawn from ancient Jewish and Christian texts, American Christians opined that a new and better world was to come. Researchers in robotics and AI adopted that perspective, merging it with the tech-nologies promised in science fiction to create a new religious move-ment: Apocalyptic AI.

Ancient apocalyptic traditions provide the ideological structures that guide 21st-century perspectives on AI: a dualistic worldview in which evil currently appears victorious but which promises that the world's troubles will be resolved in a transcendent future redemption that abolishes evil in the transformation of the world and humanity. Biblical texts in the Jewish tradition and the Christian tradition agree on this basic expectation for the coming world. Having previously articulated this in depth, I will only summarize it here.[40]

The ancient apocalypses ('unveilings' or 'revelations' of a divine plan) were composed from the second century BCE through the first century CE. Although related modes of discourse trace back centuries, such as to problems with the Assyrian empire in the Book of Isaiah (8th c. BCE), the biblical texts that address these are prophetic oracles and hopes for redemption, not proper apocalypses. God's claim in Isaiah (65:17) that 'I am about to create a new heavens and a new earth' is metaphorical, but it is fully realized in 1 Enoch (93:16) 'And the first heaven shall depart and

---

[38] See Noble, *Religion of Technology*, 24–34; Ali, '"White Crisis" and/as "Existential Risk."'
[39] See Schoepflin, 'Apocalypticism in an Age of Science', 427; Boyer, *When Time Shall Be No More*, 157–62.
[40] See Geraci, *Apocalyptic AI*, 140–6 and *Apocalyptic AI*, 14–21.

pass away, and a new heaven shall appear' and in the Book of Revelation (21:1).[41] The true apocalyptic texts trace to Greek rule and Roman rule over Israel.[42] It is these texts which unfold the full logic of cosmic and human transformation that is the hallmark of apocalyptic salvation.

These apocalyptic texts describe a dualistic worldview in which believers suffer from alienation. Good and evil are at war in the world, and the faithful suffer because evil reigns. 'Apocalyptic discourse is dualistic temporally, spatially, and socially. It divides this world from the world to come, earth from heaven, and us from them—dwellers in heaven from dwellers on earth, children of light from children of darkness'.[43] This dualistic perspective aligns everything in the cosmos: heaven and earth, saints and sinners, angels and devils, souls and bodies (it is important to note, however, that for most apocalyptic believers, it is not bodies in general but the specific bodies inhabited by sinful humanity which are problematic; this matters for their expectations of redemption).

Having structured the world under the opposition of good and evil, it is not hard for apocalyptic communities to see evil as governing the world. Such interpretations are predictable thanks to human psychology: we have a cognitive bias towards the negative.[44] Not everyone with a dualistic view of the cosmos is apocalyptic: in theory, the forces of good could be winning. So the alienation experienced by those who see the world as a domain of evil is a crucial aspect of apocalyptic thinking—it is this that will demand cosmic restitution. 'Apocalyptic is a language of crisis', but the form of such crises can vary considerably.[45] Sometimes, political and economic disenfranchisement produces apocalyptic thinking, such as we see in 1 Enoch (103:11): 'We hoped to be the head and have become the tail: We have toiled laboriously and had no satisfaction in our toil; and we have become the food of the sinners and the unrighteous, and they have laid their yoke heavily upon us'. But even those in political power can

---

[41] On the metaphorical nature of early prophetic oracles, see Collins, 'From Prophecy to Apocalypticism', 141.

[42] Apocalyptic texts composed under Greek rule include Daniel, 2 Maccabees, early compositions of 1 Enoch. Those composed under Roman rule include 2 Baruch, 4 Ezra, Apocalypse of Abraham, later compositions of 1 Enoch, parts of the Dead Sea Scrolls, Chapter 13 of the Gospel of Mark, apocalyptic elements in Paul's letters, and Revelation.

[43] Meeks, 'Apocalyptic Discourse and Strategies of Goodness', 463.

[44] See Rozin and Royzman, 'Negativity Bias, Negativity Dominance, and Contagion'.

[45] Russell, *The Method and Message of Jewish Apocalypticism*, 6. On variation, see Geraci, *Apocalyptic AI*, 16, 26, 171n23.

find themselves staring at a grim and dark world. In the ancient context of Greek and Roman rule, even well-off Jews faced a constant existential crisis in terms of their covenant with God and the moral and political questions raised by foreign rule.[46] Similarly, religious evil can produce alienation as groups debate appropriate behaviours or beliefs. Death is the most visible sign of evil's reign; in the face of it, apocalyptic thinkers anticipate an immortal future.

Despite their experience of alienation, apocalyptic believers are fundamentally optimistic.[47] This optimism reflects typical psychological responses to negativity: we respond 'to negative events with short-term mobilization and long-term minimization'.[48] That is, we react strongly and quickly to negative news, but we reduce our level of concern over time. In apocalyptic perspectives, this means that the immediacy of crisis recedes as believers anticipate their future redemption. For this reason, Hollywood films labeled 'apocalyptic' are more correctly 'eschatological'. They deal with the end of times, but not the inauguration of new times that apocalyptic believers expect will solve their alienation. In Jewish and Christian apocalyptic thinking, the end of this world is but the precursor to the beginning of a new and better world. This new world is more than simply a better world, it is one that resolves the fundamental dualism (and hence, alienation) of the old.[49] Ancient Jews and Christians both anticipated the end of the world:

> The world hasteth to pass away. (4 Ezra 4:26)
>
> The power of creation is already exhausted, and the coming of the times is very near and has passed by. (2 Baruch 85:10)
>
> Truly I say to you, that this generation shall not pass, 'til all these things be done. (Mark 13:30)

---

[46] See Horsley, *Jesus and the Spiral of Violence*, 4, 128–29; Collins, *From Prophecy to Apocalypticism*, 133, 147.

[47] See Meeks, 'Apocalyptic Discourse and Strategies of Goodness;' Searle, 'The Future of Millennial Studies'.

[48] Taylor, 'Asymmetrical Effects of Positive and Negative Events', 67. Taylor notes that people resist negative emotions (ibid., 74), which seems relevant to the optimistic gloss that apocalypticism places on its negative evaluation of contemporaneous world conditions.

[49] On apocalypticism and utopia, see Collins, *From Prophecy to Apocalypticism*. On the resolution of cosmic dualism, see Bull, *Seeing Things Hidden*, 80.

But the apocalyptic communities look forward to the end rather than fearing it; they expect that the end of this time is merely the onset of an eternal paradise. John of Patmos offers the classic statement of this in Revelation (21:1–2): 'I saw a new heaven and a new earth; for the first heaven and the first earth had passed away … I saw the Holy City, the new Jerusalem, coming down out of heaven from God'. Apocalypticism, as a social movement grounded in the worldview of apocalyptic texts, looks forward to a glorious new world, a cosmic transformation that will rectify the world's ills.[50]

To live in the glorious world to come requires the salvation of human beings. In particular, human beings will acquire new, angelic bodies. The pre-apocalyptic Hebrew traditions promised longer life in their prophetic oracles, such as the one hundred-year-old man who would be considered young (Isaiah 65:20), but this vision remains ensconced within the expectations of a world unchanged in its essence. Lacking a worldview in which the body and world are inherently problematic, it is possible for Isaiah to imagine a utopian earthly kingdom. But for the apocalyptic mind, this would not truly answer to the problem of cosmic dualism. Instead, human beings will undergo a transformation that enables them to live in the new world. Paul writes that 'flesh and blood cannot inherit the Kingdom' and so 'we will not all die, but will be changed' (1 Corinthians 15: 50–51). He goes on to argue that 'what is sown is perishable, what is raised is imperishable. It is sown in dishonor, it is raised in glory. It is sown in weakness, it is raised in power. It is sown a physical body, it is raised a spiritual body' (1 Corinthians 15: 42–44). This spiritual body belongs in the new kingdom of God, echoing previous Jewish literature, such as 2 Baruch, in which we read that the saved 'shall be glorified in changes, and the form of their face shall be turned into the light of their beauty, that they may be able to acquire and receive the world which does not die, which is promised to them' (51: 3–4) and 'they shall be made equal to the stars' (51:9). In the reign of good triumphant, human beings will experience an apocalyptic transformation that aligns them—and their bodies—with the glorious new kingdom promised by their god.

---

[50] Some scholars (e.g. Webb, 'Apocalyptic') argue that apocalypticism should not be widely applied to social movements. In *The Apocalyptic Imagination*, Collins argues that such criticisms create more problems than they solve and defends the common academic usage of apocalypticism to refer to social movements.

Such apocalyptic dreams essentially disappeared from Judaism and faded into the background of Christianity after the Roman Empire converted to that faith; but they re-emerged periodically throughout Christian history. As noted above, the medieval era capitalized on apocalyptic thinking as a strategy in the justification of both military occupation (especially with regard to the Middle East) and technoscientific progress. As Europe transitioned into colonial modernity, apocalypticism was once more on the cultural horizon. Christopher Columbus, for example, carried apocalyptic expectations with him on his voyages of conquest.[51]

Given the impact of apocalyptic justifications for Europe's early colonial enterprise and the flight of radical Christians from Europe to America, it should be no surprise that American Christianity retained a strong tradition of apocalyptic thinking. From Jonathan Edwards to William Miller, American Christianity has included popular, outspoken leaders who believed that the Kingdom of God was imminent. By the 19th century, Protestant Christians with an apocalyptic mindset believed 'religion that mattered ... was religion that brought the symbolic squarely into historical time'.[52] In the 20th century, similar ideas were held by theologians in the mainstream, such as Billy Graham, and on the fringe, such as Bob Jones and David Khoresh.

The strong commitment to scientific salvation apparent in Apocalyptic AI is thus a culmination of the transferal of apocalyptic salvation from the spiritual to the material. Initially, religion remained in the foreground, including in American beliefs that modern technologies (e.g. locomotives) indicated that the end of historical time was upon us.[53] As part of this transition, evangelical Protestant Christians made lemonade out of the 20th century's bitterest lemons. Billy Graham, for example, believed that nuclear war is predicted as the fire from heaven in 2 Peter (3: 10–13) and would herald the return of Jesus to Earth.[54] While scientists feared a world-ending catastrophe, vividly illustrated by the Doomsday Clock,

---

[51] Watts, 'Prophecy and Discovery'.

[52] Albanese, *America*, 168. Albanese refers specifically to dispensationalist Christians, whom I here situate within the category of apocalyptic.

[53] Albanese, *America*, 477.

[54] McLaughlin, *Modern Revivalism*, 140 cited in Apel, 'The Lost World of Billy Graham', 143. Admittedly, many ordinary fundamentalist Christians showed aversion to the prospect of annihilation though simultaneously they expressed little interest in opposing it; see Strozier and Simich, 'Christian Fundamentalism and Nuclear Threat', 87–9, 93.

some Christians anticipated the new world and their theological vindication. This complex soup of optimistic apocalypticism, political anxiety, science fiction futurism, and looming technoscientific calamity nourished the emergence of a new technologically fueled and scientifically justified religion: Apocalyptic AI, brought to 21st prominence by its iron horsemen.

The Iron Horsemen: Robot, Cyborg, Avatar, and Prophet

A wide variety of authors and thinkers have contributed to the Apocalyptic AI movement, but here I limit myself to a traditional number of four, each of whom contributes to the development of Apocalyptic AI in science, engineering, and pop culture: Hans Moravec, Kevin Warwick, Philip Rosedale, and Ray Kurzweil.[55] Their status as scientists and technical innovators and (for three of them) their high-profile publication of pop science books give them social credibility that 20th-century science fiction authors and transhumanist advocates generally lacked. Whereas Arthur C. Clarke imagined a future of minds transferring from body to body or from body to computer, the inspiration he offers did not capture the imagination of policymakers, businesspeople, and news media the way the iron horsemen have in 21st-century America.[56] His futuristic vision could inspire, but it lacked the seriousness and the weight carried by successful research careers, pop science books, and entrepreneurial success. Each in his own way, Moravec, Warwick, Rosedale, and Kurzweil provide the scientific and technological justification for a transition from science fiction to science futurism.

---

[55] A variety of other authors have aligned themselves with major aspects of the Apocalyptic AI movement. For examples, see Minsky, 'Will Robots Inherit the Earth?' Crevier, AI; Vinge, 'Technological Singularity;' de Garis, The Artilect War; the AI designers described in Kelly, 'Nerd Theology', 389; Newell, 'Fairy Tales'. Scholars have noted that many of the essential characteristics of Apocalyptic AI thinking percolate throughout scientific and entrepreneurial communities; see Helmreich, Silicon Second Nature, 83–5, 193; Wertheim, The Pearly Gates of Cyberspace; Lanier, You Are Not a Gadget, 25. Some luminaries have argued about the likelihood of Apocalyptic AI claims and have found themselves closely followed by pop media; e.g. Wakefield, 'Elon Musk Reveals Brain-Hacking Plans;' BBC News, 'Elon Musk and Jack Ma Disagree about AI's Threat'.

[56] It should be noted that Clarke did make contributions to scientific theory, particularly by noting the possibility of geo-synchronous orbit. Nevertheless, he has less obvious credibility in science and engineering than the iron horsemen.

Hans Moravec, formerly of the Carnegie Mellon University Robotics Institute and a seminal researcher in mobile robotics, made the most innovative contributions to the movement: he proposed a scientific basis for transcendent machines and mind uploading, and he offered a sustained argument for the transition of biological into robotic life. First in a 1978 essay in *Analog* and subsequently in his books *Mind Children: The Future of Robot and Human Intelligence* (1988) and *Robot: Mere Machine to Transcendent Mind* (1999), Moravec argues that progress in computing technologies indicates the imminent arrival of intelligent machines and thus makes transcendentally intelligent machines inevitable.[57] Human beings cannot compare with the thinking speeds and intellectual power of robots that learn by download, never forget, and can process computation at near the speed of light.[58] Our own transition into robots is 'inevitable and desirable'[59], and this evolution into machines will produce an unstoppable 'Mind Fire' in which computation spreads throughout the universe.[60] In the final analysis, it is Moravec who established the critical components of Apocalyptic AI.

Borrowing many ideas from Moravec, Kevin Warwick proposes that merging with machines will be vital to preserve humanity in a competitive environment. Warwick, a roboticist in the UK, agrees that progress in AI and robotics makes intelligent robots inevitable and fears that such robots might find little use for humanity.[61] Although Warwick disputes the claim that we will all transfer our minds into immortal robot bodies,[62] he anticipates a radical cyborg future where human beings transcend their evolutionary limits. Cyborgs will join AIs in a virtual world of wireless Internet connection and will develop tremendous physical and mental capabilities; in fact, Warwick believes that only by becoming cyborgs will human beings remain relevant in the future world.[63] Although

[57] The first appearance occurs in Moravec, 'Today's Computers, Intelligent Machines and Our Future'. See, particularly, Moravec, *Robot*, 13.
[58] Moravec, *Mind Children*, 55–6; *Robot*, 55. This position is echoed in Kurzweil, *The Age of Spiritual Machines*, 4; *The Singularity is Near*, 8–9; Warwick 2004, 178; Minsky, 'Will Robots Inherit the Earth?'
[59] Quoted in Chaudry, 'Valley to Bill Joy'.
[60] Moravec, *Robot*, 12–4, 163–7.
[61] Warwick, *March of the Machines*, 21–7.
[62] Warwick, *March of the Machines*, 180–1.
[63] Warwick, *March of the Machines*, 266–8; 'Cyborg Morals, Cyborg Values, Cyborg Ethics', 133; Warwick, *I, Cyborg*, 151, 295, 298–9, 304.

Warwick is not as radical as Moravec in his expectations, he nevertheless envisions a transformed world in which machine intelligence (whether as AI or as cyborg implanted humanity) reconfigures the nature of life itself. He documents his faith in technological progress in *March of the Machines* ([1997] 2004) and his own experiments with implants in *I, Cyborg* (2002). Although his research agenda leveled off in the 21st century, he shifted his publication strategies to philosophical journals where he could retread the ground of his prior empirical work and share his ideas with a new audience.[64]

Because he opposes the most radical proposals of mind uploading, Warwick is important less for his contributions to the public perception of AI and robotics than for his advocacy of human–machine hybrids and the possibility that this could be one stage in a longer historical arc. Many apocalyptic texts structure the end of time as a two-stage process, one where partial fulfillment in the future launches the elect towards an ultimate salvation in the final cosmic renewal. Warwick's public advocacy for cyborg technologies and his impressive media outreach served the overall apocalyptic agenda by popularizing the idea that human beings could compete in a coming machine universe by merging with our technology. His vision stops short of Moravec's and Kurzweil's (and thus might be more realistic), but it still relies upon a radical break between the present and the future and also on the same value system that serves the more radical futurists.

Philip Rosedale avers that he is not an avid reader of futurist authors;[65] and yet as the creator of the virtual worlds *Second Life* and *High Fidelity* he has collaborated with futurists and has inscribed Apocalyptic AI values into the technology. Rosedale is an entrepreneur and an engineer, limiting his public writing to Twitter. We get a glimpse (perhaps through a mirror or a scanner darkly) into his thoughts through his interviews with journalists and by noting the networks he has established with other tech leaders. Rosedale and *Second Life* were featured in Ray Kurzweil's documentary, *Transcendent Man*, and Linden Lab (the company founded by Rosedale) invited Kurzweil to give the keynote presentation at the *Second Life* Community Convention in 2009. Rosedale acknowledges that 'the

---

[64] For example, see Warwick, 'Superhuman Enhancements via Implants'.
[65] Rosedale, email correspondence with the author, April 7, 2011.

experience of creating *Second Life* had at least a lot of us convinced that it was inevitable that similar approaches would allow us to simulate the brain (or something like it) and therefore got us excited about the general area' of futurism and mind uploading.[66] Journalist Tim Guest reports that Rosedale told him 'there's a reasonable argument that we'll be able to' leave our bodies behind and upload our minds into virtual reality.[67] Similarly, in an interview with Melinda Byerley, Rosedale states that 'we are certainly not evolving as quickly as our computers'.[68] This comfortably aligns him with the other horsemen, for whom it is precisely the evolution of computers that brings about the new world. Later, in the same interview, he says, 'we're in exponential time scales. So hopefully we all live to the end of days here'.[69]

Rosedale's companies create immersive online experiences. Participants create virtual avatars through which they interact in the virtual environment. So while Rosedale makes few concrete promises about the future compared to the other horsemen, his employment is arguably more engaged with Apocalyptic AI concepts than the labors of the others. Moravec's industry work involved factory robots. Kurzweil's natural language processing work with Google automates email replies. Warwick's cyborg efforts didn't come anywhere close to producing a real merging of human and machine intellect. But Rosedale's online worlds, so reminiscent of the science fiction that inspired him, provide real and meaningful experiences of possible apocalyptic futures. There exists avid use of virtual environments and videogames as transhumanist playgrounds, and many transhumanists believe that virtual worlds and games encourage non-believers towards transhumanism.[70] As a result, Rosedale's online platforms are clearly important contributions to the futurist expectation of radical cosmic transformation.

Our fourth horseman is Ray Kurzweil, whose most significant contributions to the Apocalyptic AI movement have been synthesizing, networking, and popularizing. A prolific and important innovator in his

---

[66] Rosedale, email correspondence with the author, April 7, 2011.
[67] Guest, *Second Lives*, 273.
[68] Byerley, 'Philip Rosedale', 40 minutes, 40 seconds.
[69] Byerley, 'Philip Rosedale', time 55 minutes, 10 seconds. Unfortunately, Byerley swiftly moves the conversation past this immensely provocative statement.
[70] See Geraci, 'Videogaming and the Transhuman Inclination;' Geraci, *Virtually Sacred*, 184–197.

own right,[71] Kurzweil sharpened the focus of Moravec's argument by providing extensive supplementary data and by producing a more compelling and more human vision of the future than Moravec's sterilized cyberspace of data computation. Kurzweil amplifies Moravec's attention to exponential progress and combines it with the catchy concept of the Singularity (popularized by mathematician and science fiction author Vernor Vinge[72]). In addition to his successful pop science books, especially *The Age of Spiritual Machines* (1999) and *The Singularity Is Near* (2005), Kurzweil has been involved in multiple documentary films; been interviewed across the American news spectrum; cofounded a for-profit educational institution (Singularity University) to promote his ideas among entrepreneurs, business leaders, and policymakers; and even authored a novel to inspire young people.[73] He has even been invited to deliver a keynote address at the Nobel Prize gatherings.[74] His outreach yields powerful dividends: he is widely considered America's preeminent futurist, and he has been far more successful than anyone else (outside of science fiction) in sharing Apocalyptic AI with the general public.

Basic search data demonstrate Kurzweil's public preeminence among the horsemen (see Table 2.1). His wide-ranging public engagement means that he sells more books and appears in more conversations about technology and our future. His books are considerably easier to access via libraries, are the only ones that can be purchased for Kindle (as of late 2019), and sell at higher frequencies than the other authors. By mid-2019, he had nearly 10,000 Twitter followers, which is a modest number (at best), but he gathered these followers *without ever actually posting a tweet*. Finally, he appears in vastly more Google searches than the others, with results returned nearly an order of magnitude greater than Moravec. Powerfully, Rosedale appears in more results than anyone but Kurzweil despite having never published a book and thus not having that as an anchor for reviews, mentions, and other possible results; his popularity is

[71] Long before joining Google's natural language processing efforts, Kurzweil gained fame as the inventor of the Kurzweil music synthesizer, optical character recognition, flatbed scanners, and the first reading device for the blind.

[72] Vinge, 'Technological Singularity'. Vinge himself borrowed the term from Vannevar Bush; see Ulam, 'Tribute to John von Neumann'.

[73] Kurzweil, *Danielle*. See Geraci, 'Popular Appeal of Apocalyptic AI' on Singularity University.

[74] Kurzweil, 'Ray Kurzweil Talks at the Nobel Prize Events'.

Table 2.1 Search data indicating relative popularity of Apocalyptic AI authors and books. WorldCat and Google data as of 4 April 2019. Twitter followers as of April 5, 2019. Amazon sales rankings are averages acquired using daily readings from April 4, 2019 to April 24, 2019 (results taken at approximately noon, Indian Standard Time, each day).[1]

	WorldCat results (approx. # libraries)[2]	Amazon softcover sales rank	Amazon hardcover sales rank	Amazon Kindle sales rank	Google search results returned[3]	Twitter followers
Moravec's *Mind Children*	850	645,761	760,230		289,000	
Moravec's *Robot*	875	612,725	1,489,810			
Warwick's *March of the Machines*	300	3,849,336	2,275,118		254,000[4]	
Warwick's *I, Cyborg*	500	1,967,056	2,181,976		68,700	
Kurzweil's *The Age of Spiritual Machines*	2000	137,430	342,590	205,917	119,000	
Kurzweil's *The Singularity Is Near*	2000	26,444	153,852	76,554	355,000[5]	
Hans Moravec					504,000	
Kevin Warwick					1.35 million	
Philip Rosedale					1.38 million	16,663
Ray Kurzweil					4.68 million	9529

[1] Data like those included in the table are an important way in which scholars in the humanities and social sciences can leverage advancing technology to better understand their subject matter. For a theoretical argument that leverages data on other AI technologies, see Bainbridge, *Cultural Science*.

[2] Because of different listing strategies used by different libraries, it can be tricky getting an exact library count from WorldCat.

[3] Google searches for book titles were made using the title in quotation marks to limit the search to more relevant results. For example: 'Age of Spiritual Machines' rather than *Age of Spiritual Machines*. Results for 'Robot' are not included, as this would include a vast number of infiltrators. Including the subtitle would skew the results in the opposite direction given that most websites mentioning, for example, *The Age of Spiritual Machines* would not include a subtitle.

[4] While all Google searches include infiltrators, this number is clearly strongly skewed by a *Magic: The Gathering* result. Obviously, this is less noticeable than the 12.1 million results returned for 'Robot', the vast majority of which have no relation to Moravec's book of that title; nevertheless, it should be kept in mind.

[5] If 'the' is removed from the title, *The Singularity Is Near* returns 432,000 results. Removing 'the' from *The Age of Spiritual Machines* returns 143,000 results.

drawn entirely from his entrepreneurial successes, his reputation for innovation, and his inspiring engagement with technology. As one person
commented on the blog *New World Notes*: 'I have to hand it to Philip,
he can certainly inspire the imagination like nobody else'.[75] Likewise,
Melinda Byerley says of him: 'while my next guest may not be as famous
as Steve Jobs is today, when the history of this age of technology is studied,
long after I and all my guests and likely you are gone, they will talk about
Philip Rosedale. I guarantee you'.[76]

The iron horsemen of the AI apocalypse—Moravec, Warwick,
Rosedale, and Kurzweil—carry the promise of radical transformation to
our increasingly networked world. They stand for robots, cyborgs, and
avatars; they prophesize a new world. Their impact on American tech
culture and entertainment has been profound, and thanks to the technological circuits that unite the world, their impact outside of the U.S. grows
by the year. Many of their ideas, largely borrowed from science fiction
in the first place, are carried to the public in science fiction. They are the
meat upon which the 21st-century hunger for science fiction feasts—as
seen, for example, in the success of online shows, such as Netflix's *Black
Mirror* and *Altered Carbon*. But simultaneously, the iron horsemen build
and write their contributions into the 21st century technological marketplace. Their influence spans the globe, producing a philosophical and religious movement that challenges traditional religions even as it borrows
its fundamental structure from them.

Engineering religion: Apocalyptic AI

In explicit and implicit ways, the iron horsemen and their allies establish a new form of religion through their commitment to cosmic transformation and a radically new future. Drawing on the cultural resources
described earlier, each of the iron horsemen helps produce a social movement through which people define the world in a dualistic fashion, experience alienation, anticipate a radical future that transforms the cosmos,

---

[75] Clara Sellers comment on Au, 'Listen: Philip Rosedale Interviewed by Linden Vet about
Building Virtual Worlds'.
[76] Byerley, 'Philip Rosedale', 0 minutes, 47 seconds.

and look forward to a new, glorified embodiment within that future. This chapter has hinted at such a perspective, dancing around the contributions and perspectives of Moravec, Warwick, Rosedale, and Kurzweil. In this section, the Apocalyptic AI worldview will be summarized and made visible: a dualistic struggle between intellect and body (mirrored in the contrast of machine and biology) leads to a sense of alienation in the difficulties faced by human intelligence and the inevitable declines that the body imposes upon it. These can be resolved only when the world is transformed from the reign of biology to that of machines, a world occupied by intelligent machines and the uploaded or cyborg minds of posthumanity.

There are doubters, those who believe that AI brings terror rather than salvation or who believe the promises of Apocalyptic AI are overblown. Naturally, there have been science fiction dystopias that focus on robots, most famously *The Terminator* franchise. But in the 21st century, technologists themselves have joined the chorus of critics. Bill Joy of Sun Microsystems was first to do so, but he has been echoed in a variety of ways by Kevin Kelly, Jaron Lanier, Stephen Hawking, and Elon Musk.[77] Hawking says that 'the development of full artificial intelligence could spell the end of the human race'.[78] Musk notes on Twitter that 'We need to be careful with A.I. Potentially more dangerous than nukes' and 'Hope we're not just the biological boot loader for digital superintelligence. Unfortunately, that is increasingly probable'.[79] Despite such luminous objections—whether out of fear that robots will destroy us or out of skepticism regarding robotic salvation—Apocalyptic AI runs rampant in tech culture. As such, the apocalyptic trajectory traced by the iron horsemen bears understanding.

A dualistic worldview is a critical component in apocalyptic ideologies, including in U.S. conversations about robotics and AI. Moravec clearly articulated a worldview in which mind/intellect/information/machine opposes brain/body/materiality/biology. He relishes the information-bearing aspect of life but recognizes that biology places direct limits on

---

[77] Joy, 'Why the Future Doesn't Need Us;' Kelly, 'The Myth of a Superhuman AI;' Lanier, 'One Half A Manifesto;' Fagella, '(All) Elon Musk Artificial Intelligence Quotes;' McMillan, 'AI Has Arrived, and That Really Worries the World's Brightest Minds'.

[78] Cellan-Jones, 'Stephen Hawking Warns Artificial Intelligence Could End Mankind'.

[79] Musk, Twitter posts of 2 August 2014 and 3 August 2014.

the very information that he considers central to the human experience. Moravec notes that there will soon be a time 'when virtually no essential human function, physical or mental, will lack an artificial counterpart'.[80] He believes that machines will soon equal humanity in their intellectual abilities. Kurzweil agrees, claiming that 'within several decades information-based technologies will encompass all human knowledge and proficiency, ultimately including the pattern-recognition powers, problem-solving skills, and emotional and moral intelligence of the human brain itself'.[81] As machine thinking improves, 'robot industries will start as conversions of existing enterprises, retaining their institutional, legal, and competitive structures. But then they will explore and exploit expanding non-traditional options, some of them very unhuman. Our artificial progeny will grow away from and beyond us'.[82]

Dualism leads to comparison: Moravec and the iron horsemen agree that machines will hold all the advantages. Both Moravec and Kurzweil take pains to describe the speed of human thought and show its inadequacy compared with that of machines and the inevitability of machines matching and exceeding human ability.[83] Of course it is plausible to critique these positions by suggesting that thought is not equivalent to computation, that processing is not identical with comprehension or intelligence, or that faster computers are not alchemically transmuted into conscious ones; but that is beside the point. It is the power of this worldview, and its impact on the world that is of relevance. Similarly, while their assumptions about brain computation are likely inaccurate and their predicted time scale already wildly off, the basic point must be ceded: computers process information much faster than human beings within the domains they are capable of processing. Meanwhile, that processing speed, memory capacity, and the domain of tractable problems improve with time, whereas human capacity in these areas does not. Finally, digital information can, in theory, be stored and faithfully transmitted indefinitely, a feature not available to human thought. Overall, when the iron horsemen think about thinking, they see computers rapidly equaling and

---

[80] Moravec, *Mind Children*, 2.
[81] Kurzweil, *The Age of Spiritual Machines*, 8.
[82] Moravec, *Robot*, 11.
[83] Moravec, *Mind Children*, 55–6; Moravec, *Robot*, 52–7; Kurzweil, *The Age of Spiritual Machines*, 102–5; Kurzweil, *The Singularity is Near*, 122–30.

then surpassing human beings because the material substrate, silicon, can accomplish things that human brains cannot. In the present, they keenly feel the disadvantages of biological life and experience alienation in their comparison with hypothetical machine intelligence.

While it is psychologically possible for human beings to live comfortably within a dualistic cosmos, apocalyptic thinkers—by their very nature—reject the dualism as problematic and as a source of discomfort and struggle. Apocalyptic AI is inherently a worldview of alienation grounded in the superiority of machines over biology. Neurological systems have inherent limits that cannot equal computer transistors' speed, the storage capacity of computers, or the quasi-immortality of digital information.

The iron horsemen decry our limited memories, our slow learning, and our eventual deaths. Our bodies are 'frail and subject to a myriad of failure modes', not to mention the cumbersome maintenance rituals they require'.[84] This frailty helps explain the disappointing limits to their performance. 'I often do have a problem with all the limitations and maintenance that my version 1.0 body requires, not to mention all the limitations of my brain', writes Kurzweil.[85] The bodies that house our minds place inherent limits on those minds' capacity to learn and to think. Alongside faster calculation and memory, Warwick notes that machines can utilize a wider range of sensors, perceive the universe in richer fashion, and communicate far faster than human beings, whose 'communication is so poor as to be embarrassing'.[86] He says that he 'could look with jealousy, as a mere human, on the capabilities of machine intelligence, both in the fairly low-powered microprocessor brains of some of our current robots and in AI's potential future'.[87] After listing a few advantages of silicon over biology, he states 'those listed are conclusions drawn from our own research and gave me a tremendous drive to do something about upgrading myself'.[88] He predicts that the machines

---

[84] Kurzweil, *The Singularity is Near*, 9 (on biological limits see also 27, 28, 29; on mortality, see 325).

[85] Kurzweil, *The Singularity is Near*, 203. Kurzweil is here reminiscent of Esfandiary's classic work, in which he writes: 'how can I not feel alienated from my own temporary self which at any moment may slip away into permanent nonexistence? How can I help feeling alienated from my own fragile vulnerable body which brings me suffering … How can I not feel alienated from the world which at any instant can forfeit my existence?' (*Up-Wingers*, 155).

[86] Warwick, 'Cyborg Morals, Cyborg Values, Cyborg Ethics', 132.

[87] Warwick, *I, Cyborg*, 61–2.

[88] Warwick, *I, Cyborg*, 62.

will outsmart us and, unless we take proactive steps, we will be subservient to them.[89]

In the starkest expression of Apocalyptic AI alienation, Moravec and his fellow horsemen align life with machines and death with biology. Bodies die, taking information with them. In conversation with *Second Life*'s embedded journalist, Wagner James Au, Rosedale confessed that 'I've been obsessed with the idea that we're mortal ... and to be stuck in a skeleton ... it's not a good outcome, not a good situation'.[90] He further alleged that 'all we have to do now is figure out how to escape death'[91] and that 'I'd love to live forever ... I think the idea that people are only supposed to live for a hundred years is really dumb'.[92] Moravec, meanwhile, bemoans 'the wanton loss of knowledge and function that is the worst aspect of personal death'.[93] Marvin Minsky, one of the founders of AI, was an early convert to Moravec's vision; he begins an essay published in *Scientific American* by asserting that 'everyone wants wisdom and wealth. Nevertheless, our health often gives out before we achieve them. To lengthen our lives, and improve our minds, in the future we will need to change our bodies and brains ... In the end, we will find ways to replace every part of the body and brain—and thus repair all the defects and flaws that make our lives so brief'.[94]

Disappointment at human limits in learning and thinking and fear of death as the obvious (and permanent) exposure of these limits produces a sense of alienation that cannot be resolved in the world as it is. A new world must come to end the dualism, the ongoing conflict that can be won only when transcendent forces refashion the world and erase the contradiction posed by our earthly lives. In Jewish and Christian apocalyptic traditions, divine providence guarantees the new world to come. AI

---

89 Warwick, *March of the Machines*, 261, 302.
90 Quoted in Au, *The Making of Second Life*, 233.
91 Au, *The Making of Second Life*, 232.
92 Au, *The Making of Second Life*, 233.
93 Moravec, *Mind Children*, 121. Kurzweil presents a more human concern over death, one where mourning has a clearer connection to love (see Hochman, 'Reinvent Yourself'). There are, of course, parallel considerations in India. Consider an early 20th-century example where the author notes that even great geniuses can learn only a few things in one lifetime and decried the loss of 'mind' as a terrible misfortune of death—though the author considers moral losses worse than intellectual ones; see Maitra, 'The Hope of Immortality', 600–1. It should come as no surprise that this early-century Indian author places his hope in immortality in a religious afterlife rather than a scientifically extended earthly life (p. 603).
94 Minsky, 'Will Robots Inherit the Earth?'

and robotics experts cannot maintain the scientific rigor of their position, however, if they turn in this direction. Instead, they look to scientific laws as transcendent guarantees of the new world. Moravec argues that evolution is 'weeding out ineffective ways of thought'.[95] This requires a discursive shift in which evolution applies outside the realm of biology and the various principles operating there, but Moravec manages this convincingly because he leverages the everyday (unscientific) usage of the term. Evolution has come to refer to almost any form of progress, which is why Kurzweil can copy Moravec and argue that 'the next stage of evolution ... is technology'.[96] Kurzweil buttresses his claim by inventing a new natural law: the law of accelerating returns, which he believes is 'not a temporary methodology. It is a basic attribute of time and chaos'.[97] Kurzweil's presumptions—of evolution qua technological progress and increasing orderliness exponentially returning more order- combine to mean that transcendent machine intelligence would be *guaranteed* by cosmic laws.

The inevitable new world alleged by the iron horsemen will follow a typical two-stage apocalyptic pattern, with earthly utopia preceding the transcendent new world. Moravec argues that advanced technology will solve many of the problems pressing in the 20th and (now) 21st centuries: controlling weather, eliminating pollution, providing a universal basic income, making nationalism and war obsolete, and allowing human beings to revert to a 'comfortable tribalism' in a 'garden of earthly delights'.[98] Kurzweil sums this up succinctly, claiming that all human needs will be fulfilled.[99] Advancements in robotics and AI, along with accompanying progress in nanotechnology and similar areas, will supposedly make this paradise come true. But as delightful as it may seem, this is but the prelude to the real transformation (it must be acknowledged that Warwick perceives something like this to be the final stage; from his perspective, the cyborg

---

[95] Moravec, *Robot*, 165. See also Moravec, *Mind Children*, 167.

[96] Kurzweil, *The Age of Spiritual Machines*, 35.

[97] Kurzweil, *The Age of Spiritual Machines*, 33. He is echoed by de Garis, who claims that the coming of intelligent machines 'is inherent in the laws of physics;' see de Garis, *The Artilect War*, 173. The premise of Kurzweil's 'law' already existed in transhumanist communities, even in very similar wording. Esfandiary, for example, states 'the rate of advance is now accelerating. Progress is faster and more global than ever' (*Up-Wingers*, 12).

[98] Moravec, *Robot*, 9–10, 134, 136–7, 143, 155.

[99] Kurzweil, *The Age of Spiritual Machines*, 2.

integration of humanity and machine will remain tied to our essential biology[100]). Drawn to its fullest conclusion, faith in the information identity of humanity and the exponential growth of computing technologies cannot be restrained by earthly limits: a complete transition to mechanical life becomes inevitable. Machine learning means that thinking can be 'freed from bondage to a mortal body'.[101]

Biology imposes limits on computation, but exponential progress in computing technologies will—according to the iron horsemen and their followers—bring about a Singularity, a Mind Fire, a cosmic transformation. Each in his own way, all of the iron horsemen predict a future of intelligent machines. For them and their followers, such machines will rapidly outgrow humanity. Rosedale conjectures that 'the most interesting of our progeny will be things that are pure computation. It'll be possible for constructs that we build in Second Life and things like it in a simulated space to actually think'.[102] Under the assumption that progress will continue more or less exponentially and along the lines indicated by Moore's Law, a machine that equals humanity would be twice as smart as a human being within a year or two.

The oft-cited 'Singularity' is the point in this curve of progress where exponential growth means astronomical results occur in the same timespan that once permitted only modest or even miniscule results. Moravec initially proposed that human-level machine intelligence will be possible by 2010 (and by 2030 in a personal computer).[103] Somewhat cagier, Warwick predicts that 'in the next ten to twenty years some machines will become more intelligent than humans'.[104] In *The Age of Spiritual Machines*, Kurzweil says 2020 (and 2025 for a personal computer); but in his subsequent *The Singularity is Near*, he gets more ambitious, projecting 2010 (and 2020 or sooner for a personal computer).[105] If this happened, we would shortly thereafter have machines far more intelligent than we. Such robots would bring about an entirely new world: a world of machine intelligence, software immortality, and an infinitely open multiverse of cyberspace. At a minimum, these predications are optimistic.

[100] Warwick wavers over the plausibility of mind uploading but believes that we can realistically integrate machine hardware into our brains to become competitive; Warwick, *March of the Machines*, 180–1, 267; Warwick, 'Cyborg Morals, Cyborg Values, Cyborg Ethics'. Warwick speaks somewhat more favorably about mind uploading in *I, Cyborg*, 70.

[101] Moravec, *Mind Children*, 4.

[102] Quoted in Au, *The Making of Second Life*, 233.

[103] Moravec, *Mind Children*, 68; Moravec, *Robot*, 59, 63.

[104] Warwick, *March of the Machines*, x.

[105] Kurzweil, *The Age of Spiritual Machines*, 103; Kurzweil, *The Singularity is Near*, 70.

A radical future supposedly awaits, one that emerges from a drastic break in historical time. Even Warwick, more modest in his predictions than Moravec or Kurzweil, believes that 'creating cyborgs cannot be regarded in any other way than as a discontinuity, a non-linearity, in evolution' and that creating cyborgs 'completely changes the order of things'.[106] Moravec calls the future world a Mind Fire, which he describes as the literal transformation of thoughtless matter into a living world of intellect. The Earth won't be merely populated by intelligent machines; it will be transformed into a planetary computer. Seeking new knowledge and new experiences, robots will depart Earth and spread throughout the universe.

This diaspora through space will be a 'physical affair ... but it will leave a subtler world, with less action and even more thought, in its ever-growing wake'.[107] Even 'boring old Earth' will be 'swallowed by cyberspace' and 'host astronomically more meaningful activity'.[108] According to Kurzweil, 'the Singularity will ultimately infuse the universe with spirit'.[109] It 'will make life more than bearable; it will make life truly meaningful'.[110] The universe will 'wake up'; the 'dumb matter and mechanisms of the universe will be transformed into exquisitely sublime forms of intelligence' and proceed asymptotically towards divinity.[111] 'Evolution does not achieve an infinite level, but as it explodes exponentially, it certainly moves in that direction. So evolution moves inexorably towards our conception of God, albeit never reaching this ideal'.[112]

Importantly, both Moravec and Kurzweil refer to this new world as being *meaningful* compared to the alleged banality of the present Earth. Our present biology constrains life; but in the future, biology will cede primacy to machines. The machine world, with its rapid calculation (of whatever such machines would find worthy of calculating), will be a richly meaningful world, not simply a more informed one.

Rosedale describes *Second Life* and future virtual realities by the same logic. He believes that consciousness can arise within the virtual world.

---

[106] Warwick, *I, Cyborg*, 296.
[107] Moravec, *Robot*, 163.
[108] Moravec, *Robot*, 167.
[109] Kurzweil, *The Singularity Is Near*, 389.
[110] Kurzweil, *The Singularity is Near*, 372.
[111] Kurzweil, *The Singularity is Near*, 21. See also O'Reilly, 'What If We're the Microbiome of the Silicon AI?'
[112] Kurzweil, *The Singularity is Near*, 476.

'It'll breathe by itself, if it's big enough. We're helping because we're going in as avatars. It's simply the fact that if the system is big enough and has enough complexity, it will emerge with all these properties. People come from out of the dust'.[113] In 2007, Rosedale predicted that by 2017, 'it'll be possible for constructs that we build in Second Life and things like it in a simulated space to actually think'.[114] Despite the lack of results by 2019, Rosedale implied his continued faith in at least some version of the apocalyptic future on Twitter, where he writes: 'it is one thing to defend the position that we will not be able to figure out how to breathe consciousness into modern AI's, but [it] is bolder to argue against the likelihood that such things will evolve on their own inside digital worlds simulated by increasingly powerful computers'.[115] The apocalyptic guarantee for a future of superintelligent machines, therefore, comes not from human ingenuity but supposedly from innate conditions of the world.

The shift from physical reality to virtual reality, heralded by science fiction and foreseen in Rosedale's entrepreneurial projects, would be a radical break in history. Moravec suggests that we could produce an infinite number of worlds and that we might be living in such a simulated environment.[116] History thus breaks down conceptually: within cyberspace, 'entire world histories ... will be resurrected'.[117] Kurzweil maintains an archive of material by and about his father in the expectation that the latter can be resurrected through computer simulation.[118] Even in Warwick's more restrained techno-apocalypse, it is plausible that cyborgs might 'give up their individuality and become mere nodes on an intelligent machine network'.[119]

Rosedale, whose very business is building immersive virtual worlds, has said of them that 'you could create a world that was both new and I think free, and maybe escapist in some regard ... The idea ... that there really wasn't an argument why a virtual world wouldn't be as real as the real world ... We ought to be able to do evolution in there. We should be able to build spaces as big as Earth in there ... I always felt certain that we

113 Quoted in Au, *The Making of Second Life*,' 234–5.
114 Au, 'In *The Terminator*, It's Skynet's Birthday'.
115 Rosedale, Twitter post of March 31, 2019.
116 Moravec, 'Pigs in Cyberspace', 20–1; Moravec, *Robot*, 168, 173.
117 Moravec, *Robot*, 167.
118 Vance, 'The Ray Kurzweil Show, Now at the Googleplex', 55.
119 Warwick, 'Cyborg Morals, Cyborg Values, Cyborg Ethics', 136.

could create such spaces, that were equally real and detailed in comparison to the real world'.[120] The compelling experiences that people have in 21st-century virtual reality worlds, pioneered by Rosedale through *Second Life* and subsequently *High Fidelity* (and perhaps even more so through videogames[121]), provide a glimpse into what such freedom can mean. Virtual world users/inhabitants take advantage of their freedom from earthly physics to create the worlds and bodies of their dreams.

Just as apocalyptic Jews and Christians expected the imminent arrival of the end times and the transformation of their bodies into angelic or glorified bodies, the iron horsemen suggest that in our post-apocalyptic world, we will take on glorious new bodies. 'Truly I tell you', states Jesus in Chapter 13 of the Gospel of Mark, 'this generation will not pass away until all these things have taken place' (Mk 13:30). Two decades later, Paul confidently states 'we will not all die, but we will all be changed' (1Cor 15:51). Unknowingly channeling this sentiment for his post-Christian techno-apocalypse, Kurzweil says 'most of the readers of this book are likely to be around to experience the Singularity'.[122] His beliefs about exponentially developing machine intelligence coupled with projected technological advances in other domains, such as nanotechnology, explain Kurzweil's confidence.

The transformation itself will reconfigure our bodies, advancing humanity from its biological roots to machine bodies. Warwick believes we will stop once we become cyborgs: 'but in doing so the very fabric of humanity itself [has] to change'.[123] Moravec and Kurzweil promise that we will shift entirely to machine bodies and, once there, likely lose interest in physical bodies at all. In one fashion or another, human beings will secure their redemption by 'uploading' their consciousness into machines. Warwick wavers over the plausibility of mind uploading but believes that we can realistically integrate machine hardware into our brains to become competitive. Moravec argues that a machine could read the pattern of conscious activity in a brain while operating upon it—copying that pattern into another machine to transfer an individual's consciousness

---

[120] Byerley, 'Philip Rosedale', time 12 minutes, 20 seconds.
[121] On the relationship between videogames and transhumanism, see Geraci, 'Video Gaming and the Transhuman Inclination'.
[122] Kurzweil, *The Singularity is Near*, 323.
[123] Warwick, *I, Cyborg*, 299.

from the now sliced up brain and body on its operating table into a powerful new robot body.[124] Apparently squeamish about slicing up human bodies in order to preserve human minds, Kurzweil suggests that we will simply use brain scanning technologies to identify the pattern of neurochemical activity and then simulate that in a new body.[125]

The iron horsemen argue that we can take on new bodies because our essential selves have nothing to do the materiality of our present existence; rather, they believe that a person's identity is determined by the informational pattern computed by our bodies, particularly our brains. Moravec disparages human bodies and defines our real selves as the pattern and the process computed by those bodies.[126] His opinion is shared by Kurzweil and Warwick.[127] Defining human personhood as a neurochemical pattern is crucial to the futuristic promises of the iron horsemen. Following a different trajectory—one that draws on Nikolai Fedorov (1829–1903) and his beliefs about archives—there are Russian 'immortalists' (often called 'Cosmists') who echo this commitment to pattern-identity: they use genomics and cloud computing to advance a global archive of human beings.[128] Western transhumanists have increasingly learned about and found common cause with Russian Cosmists, and the agreement appears vibrantly in their shared vision of human beings as information patterns.

Accepting the pattern-identity hypothesis not only justifies the possibility of mind uploading, it ensures human immortality, the ultimate promise of apocalyptic redemption. Moravec alleges that 'if the process is preserved, I am preserved. The rest is mere jelly'.[129] The religious implications of pattern-identity are clear in Kurzweil's view of the future, a future in which machine bodies replace biological bodies:

> Actually there won't be mortality by the end of the twenty-first century ... Up until now, our mortality was tied to the longevity of our *hardware*. When the hardware crashed, that was it ... As we cross the

---

[124] Moravec, 'Todays Computers, Intelligent Machines and Our Future;' Moravec, *Mind Children*, 108–10; Moravec, *Robot*, 142–3.

[125] Kurzweil, *The Age of Spiritual Machines*, 124–6; Kurzweil, *The Singularity is Near*, 198–202.

[126] Moravec, *Mind Children*, 117.

[127] Kurzweil, *The Age of Spiritual Machines*, 54–5; Kurzweil, *The Singularity is Near*, 383–4; Warwick, 104.

[128] Bernstein, 'Life Unlimited'.

[129] Moravec, *Mind Children*, 117.

divide to instantiate ourselves into our computational technology, our identity will be based on our evolving mind file. *We will be software, not hardware* ... As software, our mortality will no longer be dependent on the survival of the computing circuitry ... we won't throw our mind file away when we periodically port ourselves to the latest, ever more capable 'personal' computer ... Our identity and survival will ultimately become independent of the hardware and its survival. Our immortality will be a matter of being sufficiently careful to make frequent backups.[130]

These immortal software entities—living in robot bodies or virtual reality—would revel in power and glory. Basing his assumptions on unending exponential growth in technological capability, Kurzweil argues that 'as we port ourselves, we will also vastly extend ourselves. Remember that $1000 of computing in 2060 will have the computational capacity of a trillion human brains. So we might as well multiply memory a trillion fold, greatly extend recognition and reasoning abilities, and plug ourselves into the pervasive wireless-communications network. While we are at it, we can add all human knowledge—as a readily accessible internal database'.[131] The apocalyptic promise of transcendence predicts (and describes as inevitable[132]) the rise of these wondrous immortal beings in the near future.

Apocalyptic AI authors describe different kinds of machine bodies, including non-humanoid bodies, humanoid bodies, and virtual bodies existing only in cyberspace. Moravec, for example, believes that humanity will forsake its four-limbed structure and take on new morphologies, including 'robot bushes'. His robot bushes have many limbs, each of which branches repeatedly until it reaches its terminus in many nanosized fingers capable of manipulating matter at the atomic scale. For such machines, 'the laws of physics will seem to melt in the face of intention and will. As with no magician that ever was, impossible things will

---

[130] Kurzweil, *The Age of Spiritual Machines*, 128–9, emphasis original. See also Moravec, *Mind Children*, 123–4; Moravec, *Robot*, 142.

[131] Kurzweil, *The Age of Spiritual Machines*, 126–8.

[132] Moravec, *Robot*, 13; Kurzweil, *The Age of Spiritual Machines*, 17–36; Kurzweil, *The Singularity is Near*, 407. Warwick describes the rise of intelligent machines as inevitable but, as has been noted, restricts his prophetic warning to that point; *March of the Machines*, 302.

simply *happen* around a robot bush'.[133] Kurzweil suggests that we will re-place parts of our bodies with nanotechnology and computer hardware, a process that will continue until ultimately there is nothing biological left.[134] Most people find this vision more compelling than the 'Moravec operation', in which the brain is sliced up and copied. 'As we learn the operating principles of the human body and brain', Kurzweil writes, 'we will soon be in a position to design vastly superior systems that will last longer and perform better, without susceptibility to breakdown, disease, and aging'.[135]

While Moravec and Kurzweil retain an interest in physical bodies, both believe that the future will make them obsolete. Cyberspace vi-sions articulated in science fiction—such as Vernor Vinge's Other Plane, William Gibson's Matrix, and Neal Stephenson's Metaverse—caught the attention of AI advocates and became real in the work of videogame and virtual world designers such as Rosedale.[136] Moravec recognizes that a sense of embodiment is necessary for human beings, who cannot long re-tain sanity during sensory deprivation, but he believes that ultimately the posthuman machines will 'streamline their interface' by reducing mental calculation of the world to machine language, forsaking our human ex-perience altogether.[137] Kurzweil does not go this far, accepting that our virtual environments and virtual bodies will provide sufficient satisfac-tion. In any case, he agrees that we will transcend our robot bodies: 'we don't always need real bodies. If we happen to be in a virtual environ-ment, then a virtual body will do just fine'.[138]

The machine bodies of the future—both robot and virtual—offer im-mortal salvation to posthuman minds. Writing within the context of his research into videogames, well-known sociologist William Bainbridge says 'I would consider a continued existence for my main [*World of Warcraft*] character, behaving as I would behave if I still lived, as a real-istic form of immortality ... Ultimately, virtual worlds may evolve into the first real afterlife, not merely critiquing religion but replacing it'.[139]

---

[133] Moravec, *Mind Children*, 107–8, emphasis original.
[134] Kurzweil, *The Age of Spiritual Machines*, 52–3; Kurzweil, *The Singularity is Near*, 384–5.
[135] Kurzweil, *The Singularity Is Near*, 302.
[136] Vinge, *True Names*; Gibson, *Neuromancer*; Stephenson, *Snow Crash*.
[137] Moravec, *Robot*, 170, 172.
[138] Kurzweil, *The Age of Spiritual Machines*, 142.
[139] Bainbridge, *The Warcraft Civilization*, 62.

Bainbridge's position depends importantly on the question of personal immortality but also on the potential for computer simulation to realize the heavenly aspirations of many religions. Moravec argues that simulating minds would be more than immortality; it would enable resurrection of the dead.[140] We could resurrect the dead by using massively powerful computers to reconstruct their personalities from known information and simulating them in cyberspace. Kurzweil has been outspoken about his desire to use such technology to bring back his father.[141]

Rosedale's production of actual virtual worlds that encourage users to explore and live online offers incentive for people to believe in the promises of uploaded immortality and resurrection. In a survey I conducted with users of *Second Life*, more than half of the respondents would 'definitely' or 'maybe' consider uploading their minds into the world if it were technologically feasible.[142] Many transhumanists believe that games automatically incline players towards transhumanist expectations and can even be used as deliberate evangelism for transhumanism.[143] As virtual world inhabitants increasingly acclimate to virtual environments and use those for work, play, and maintaining relationships, many will find themselves drawn towards the apocalyptic visions of Moravec and Kurzweil. While Warwick's and Rosedale's commitment to the apocalyptic vision may be only partial, they are nevertheless key players (especially Rosedale) in the economy of faith that builds and sustains Apocalyptic AI.

Tech advocates, from scientists to entrepreneurs, frequently revel in apocalyptic dreams. They believe that the future will be radically different and bring a glorious transcendence of humanity. One employee at Google, for example, reported to me that among the many internal mailing lists, many Google employees participate in one covering the Singularity and transhumanist aspirations. Apocalyptic AI, he explained, is one of the two dominant discourses about AI in the company.[144] The alienation experienced by the iron horsemen, brought on by our biological limits,

---

[140] Moravec, *Mind Children*, 122–4; Moravec, *Robot*, 142, 173.

[141] For examples, see Berman, 'Futurist Ray Kurzweil Says He Can Bring His Dead Father Back to Life Through a Computer Avatar;' Rennie, 'The Immortal Ambitions of Ray Kurzweil;' Vance, 'The Ray Kurzweil Show, Now at the Googleplex'.

[142] See Geraci, *Apocalyptic AI*, 89.

[143] Geraci, 'Video Games and the Transhuman Inclination', 747–51.

[144] The other major perspective on AI evidently in circulation among Google's employees is concern for the ethical implications of AI in surveillance, war, and other immediate technologies.

will supposedly be overcome in a machine world of superintelligence of godlike AI and uploaded human minds. Regardless of its likelihood, this cosmic trajectory circulates in tech communities and in pop culture.[145] Even government policymakers have begun reflecting on the Singularity.[146] In the opening decades of the 21st century, Apocalyptic AI joined the U.S. cultural mainstream.

Conclusion

Transhumanist thinking arose out of Christian speculation about technology. Medieval and early modern Christians believed that technology serves a divine purpose of perfecting humankind and the world, ultimately bringing about a new world. Their position remains influential within Christianity, but its significance lies increasingly in the fact that it undergirds secular transhumanism and helps ensure 21st-century confidence in technological salvation. From Fedorov and Huxley to Moravec and Kurzweil, the 20th century witnessed ever more powerful transhumanist claims.

The labors of the iron horsemen have not gone unnoticed by scientists. In fact, interest in brain–machine interfaces and human transcendence now occupies federally funded research time in the U.S. For example, Zhang and collaborators created a new mechanism for bioelectronic interfaces in 2019, one that could be used to 'store and transfer memories' to a computer.[147] Such work aims towards compensating for memory loss or other degenerative brain conditions, but the potential for such cyborg technologies is not lost on transhumanists. Policy science also attends to claims about transformative superintelligence—leading scholars in international affairs take seriously the claim that researchers will soon devise intelligent machines and that these challenge our understanding of human (and non-human) rights.[148]

---

[145] See Lanier, 'One Half A Manifesto;' Vance, 'The Ray Kurzweil Show, Now at the Googleplex'.

[146] For example, see Joint Economic Committee, 'Nanotechnology'.

[147] See Zhang, et al., 'Perovskite Nickelates as Bio-Electronic Interfaces;' Robitzsky, 'Scientists Say New Quantum Material Could 'Download Your Brain;'

[148] Livingston and Risse, 'The Future Impact of Artificial Intelligence on Humans and Human Rights'.

The futuristic ideas of the iron horsemen are genuinely important to contemporary culture, from entertainment to religion to technology. They structure our view of technology and our technological future. They also structure the future of traditional religious ideas, many of which may find themselves adopted and adapted in secular culture.[149] The iron horsemen ride with an apocalyptic worldview that promises an end to alienation in a glorified world to come. Apocalypticism depends on a dualistic worldview in which alienation can be resolved only through the arrival of a new world that humanity will occupy in new bodies. Just like ancient apocalyptic believers, Moravec, Warwick, Rosedale, and Kurzweil anticipate a new world arriving in their lifetimes. A cosmic transformation from biological life to machine life as cyborgs, robots, and virtual world avatars will overcome the difficulties that our bodies impose on our minds. The iron horsemen and the public they represent look forward to a world of transcendent machine intelligence and are building a religious worldview that has rapidly spread across U.S. pop culture and tech culture and which seeks to displace, or at least reconfigure, traditional views.

Apocalyptic AI emerged in the U.S. and was a product of many centuries of theological and political interpretation, but it bears upon the entire world. Twenty-first-century scientific thinking draws upon the futurist speculation of Apocalyptic AI even as world cultures reinterpret those beliefs. In the next chapter, we will consider how Apocalyptic AI grew and then the logic of its transition into Indian science and public life. The iron horsemen have a cosmic vision, but the shift of transhumanist speculation into India opens the door for new models of futurism that can impact worldwide approaches to technology.

---

[149] It is not clear that transhumanist religion can or will replace traditional religions. On this, see Kostick, et al., 'Engineering Eden'.

3

Bearers of the Apocalypse

Horses, Robots, and the Digital Future of India

Introduction

The iron horsemen ride digital horses. They ride them across global tech culture, including into Bangalore. As we saw in the previous chapter, a Christian devotion to apocalyptic eschatology—the radical inauguration of a glorious new world—has percolated from the biblical Book of Revelation to pop culture in the West. Today, that devotion radiates from the scientists and engineers who seek the transmutation of biological life into cyborgs, robots, and immortal online avatars. Whether or not they will succeed is beside the point; they represent a shift in the religious worldviews of modernity and their vision of artificial intelligence (AI) influences technological development, policymaking, and public life. The comparative absence of similar ideas in late 20th- and early 21st-century India shows that those claims do not map as easily onto other cultural contexts. And yet, there are bearers of Apocalyptic AI in India today.

This chapter explores their contributions with analogy to the Indian kings who established dominion and inaugurated new worlds through the famed *Ashvamedha* (horse sacrifice). The new religious movements that wed traditional Indian religious ideas with faith in technological transcendence have few champions in India, but their number grows, thanks to a new cultural and technological environment. The rise of more robust Indian science fiction and the increasing distribution of the Internet through smartphones encourage transhumanist speculation. As Apocalyptic AI catches the imagination of the Indian public, that public must decide which traditions, practices, or ideas to sacrifice.

Futures of Artificial Intelligence. Robert M Geraci, Oxford University Press. © Oxford University Press 2022.
DOI: 10.1093/oso/9788194831679.003.0004

The local context matters for ideas about technology just as it matters for technologies themselves. Thai philosopher Soraj Hongladarom has pointed towards the importance of local value systems for the deployment of technology, arguing that a nation's values can shape technological development.[1] Hongladarom notes, like Kurzweil, that nanotechnology 'is poised to change the very constitution of the body itself' and, rather differently, that 'what is needed, in short, is that the goals, agenda and contexts of science and technology should essentially belong to the local culture'.[2] In his context, taking Buddhism seriously would mean that Thai nanotechnology might philosophically reject the western notions of individual selfhood and politically 'pay particular attention to the role of compassion, commiseration and on taking concrete action to help others'[3]—seemingly a far cry from the immortalist conjectures of Apocalyptic AI. While advocates of Apocalyptic AI presume the cultural ubiquity of their values, Hongladarom suggests that technologies must account for different values in different contexts. This means that the ideas we have about our technologies are subject to change as they travel the globe, a process that can recursively affect the cultures of origin.

The Apocalyptic AI movement encourages people to see a glorious new world on the digital horizon. For good or ill, the iron horsemen and their allies hope that Apocalyptic AI will direct global priorities. In India, however, a number of obstacles stand in the way of such a trajectory—only through sacrifice can the world be rendered anew according to Indian traditions, and thus the transhumanist promises of Apocalyptic AI may ride across the Indian landscape but no transformation will be possible without changes to the western agenda. The rising voices of Indian transhumanism will encourage, and perhaps force, western transhumanists to reckon with the religious aspects of their belief system and oppressive social structures that benefit that system; simultaneously, Indians will forge a new vision of technology, one that looks to the future rather than the past.

---

[1] Honglardarom, 'Nanotechnology, Development and Buddhist Values', 100.
[2] Honglardarom, 'Nanotechnology, Development and Buddhist Values', 103.
[3] Honglardarom, 'Nanotechnology, Development and Buddhist Values', 104, 105.

Transhumanism rampant

It would be possible to trace the breadth of transhumanism through U.S. culture; but for the purposes of this book, only the impact of Apocalyptic AI is critical. Our intent is to evaluate how religious and cultural environments in India and in the West, particularly the U.S., might impact the reception, use, and future progress of digital technologies. Apocalyptic scenarios of salvation or damnation now form the core of thinking about AI across many cultural domains. Their influence is such that noted pop science editor/publisher John Brockman begins his introduction to *Possible Minds* with 'artificial intelligence is today's story—the story behind all other stories. It is the Second Coming and the Apocalypse at the same time: good AI versus evil AI'.[4] We know from the previous chapter that separating the Second Coming from the Apocalypse is a misuse of terms, and it is clear that the words are metaphorical (rather than Christian), but we can recognize the spirit behind Brockman's phrasing. The risks of AI and the salvation promised by the iron horsemen are built into contemporary AI, as was apparent in the previous chapter, and, as Brockman's book demonstrates, such beliefs have direct influence on science, technology, entertainment, and interpretations of what it means to be human.

The pivotal role of Moravec's *Mind Children* can be seen by looking at books published shortly before and after it. In the late 1980s, there were a precious few authors such as astronomer John Barrow and physicist Frank Tipler who—following the same trajectories as Moravec—argued that machines could spread throughout the universe as descendants of humankind.[5] Curiously, however, they do not go as far in their speculations. They define a human soul as the 'pattern' or 'program' running on a body and note that 'in principle' such a pattern could be 'stored' on a computer but then seem to reject the idea that a machine could fully emulate the 'very special hardware' of the body.[6] Other sources show a stark contrast that reveals the importance of Moravec's work. Eric Drexler's seminal work on nanotechnology, *Engines of Creation* (published two

---

[4] Brockman, 'Introduction', xv.
[5] Barrow and Tipler, *The Anthropic Cosmological Principle*, 615.
[6] Barrow and Tipler, *The Anthropic Cosmological Principle*, 659.

years prior to *Mind Children*), dances around transhumanism and life extension, offering a vision of near-paradise, but the book does not articulate a true posthuman future or a cosmic transformation.[7] In his history of AI published five years after *Mind Children*, however, Daniel Crevier notes that 'the Moravec operation' indicates the potential compatibility between religious views of salvation and scientific views of materialism.[8]

The importance of Apocalyptic AI is such that the iron horsemen became central to the entire scope of scientific futurism, not just books about AI. For example, renowned physicist and science popularizer Michio Kaku describes mind uploading and posthuman evolution as a key possibility for the 21st century.[9] Similarly, Tipler's widely read *Physics of Immortality* borrows from Moravec in defending his claim that 'science can now offer *precisely* the consolations in facing death that religion once offered. Religion is now part of science.'[10] The potential for mind uploading and its religious implications were thus circulating in conversations about AI soon after, and presumably owing thanks to, Moravec's first book.

Interest in mind uploading, the Singularity, and digital salvation became common in U.S. tech culture. Early contributions and support came from luminaries such as Marvin Minsky (AI), Martin Rees (cosmology), Warren Robinett (videogaming; software and hardware design), Danny Hillis (software and hardware design), Mark Pesce (VRML), and Sam Harris (neuroscience).[11] According to virtual reality pioneer Jaron

---

[7] See Drexler, *Engines of Creation*.

[8] Crevier, *AI*, 279–80.

[9] Kaku, *Visions*, 100, 116, 118, 202–3.

[10] Tipler, *The Physics of Immortality*, 339, emphasis original. Tipler alleges that his own ideas were independently and simultaneously formed; see *The Physics of Immortality*, 16–7. In this, he refers to his earlier article, 'The Omega Point as *Eschaton*'. That article was published in 1989, more than a decade after Moravec's 'Today's Computers, Intelligent Machines and Our Future' and just one year after *Mind Children*, but it is important to note that Tipler might not have seen either (particularly the relatively obscure essay). He could have been influenced by the same science fiction authors, cyberneticists/AI theorists, and philosophers that interested Moravec. Certainly, Moravec has priority in these ideas, but it is possible that Tipler is not relying on him. Tipler agrees that a human being is fundamentally a computer program ('The Omega Point as *Eschaton*', 222–3), that machine intelligence will replace human beings (ibid., 245), and that computer simulation will enable the resurrection of the dead (ibid., 246–7). Tipler even elucidates the simulation hypothesis, though he turns away from the idea that we might be living in such a simulation ('The Omega Point as *Eschaton*', 241–2).

[11] Minsky, 'Will Robots Inherit the Earth?;' Rees, 'Organic Intelligence Has No Long-Term Future;' Robinett, 'The Consequences of Fully Understanding the Brain', 169–70; Hillis, A Time of Transition/The Human Connection; Pesce, 'True Magic', 226, 237–8; Harris, 'Can We Avoid a Digital Apocalypse?' In *What To Think about Machines That Think*, John Brockman collects

Lanier, faith in the Singularity has become the de facto religious commitment of Silicon Valley.[12] Similarly, any selection of writings about the role of AI in our future will inevitably mention one of the iron horsemen (usually Kurzweil) and their futuristic speculation—though not necessarily with credulity or commitment to such outcomes. This includes writings by journalists and by tech entrepreneurs.[13] Even the most sober-minded scientists feel compelled to note the transhumanist speculation of their colleagues.[14] As Kurzweil gained popularity in the public arena, the main themes of Apocalyptic AI were most closely associated with him and became part of his role in the entrepreneurial sector. In 2009, Kurzweil and Peter Diamandis established Singularity University, an institution dedicated to bringing Kurzweil's gospel of exponential growth to policymakers and business leaders.[15]

Every year, new scientists and entrepreneurs enter the apocalyptic discourse. Overall, the term 'transhumanist' was 160 times more likely to appear in a published book in 2008 than in 1982.[16] Books on artificial intelligence are not just prone to engaging with transhumanist discourse, it has become functionally necessary for them to do so. Scientists and engineers, perhaps out of good-natured intellectual fun but also genuine intellectual interest, vigorously debate the various claims articulated by the iron horsemen: the Singularity, transcendent machines, uploaded immortality, resurrection through digital simulation, the cosmic spread of machine intelligence, and infinite virtual worlds.

Moravec's simulation hypothesis is an increasingly popular topic of discussion that has emerged from the Apocalyptic AI vision. Science fiction authors, such as Frederick Pohl in 1955,[17] previously described people living in artificial worlds but as with so much in the Apocalyptic

brief essays by a host of scientists, scholars, and others; many of these include reference to the Singularity, mind uploading, and similar concerns (a few are cited here). Other, less well-known figures in AI also joined this chorus; for examples, see Levy, *Robots Unlimited*; de Garis, *The Artilect War*.

[12] Lanier, *You Are Not a Gadget*, 25.
[13] For example, Markoff, *Machines of Loving Grace*, 84–5, 116–25.
[14] For examples, Nourbakhsh, *Robot Futures*, 106–7; Nourbakhsh and Keating, *AI and Humanity*, 37, 67; Kaplan, *Artificial Intelligence*, 138–55; Hussain, *The Sentient Machine*, 36–7; Perkowitz, *Digital People*, 186, 209, 214; Wallach and Allen, *Moral Machines*, 190–4.
[15] On Singularity University, see Geraci, 'The Popular Appeal of Apocalyptic AI', 1016–7.
[16] Google nGram viewer data; English data only.
[17] Pohl, 'The Tunnel Under the World'.

AI framework, it was Moravec who took these speculations seriously and provided them with a meaningful scientific basis. In short, Moravec argued that (1) in any universe where intelligent life might evolve, these lifeforms would develop digital computation, (2) through that they would learn to create highly, even perfectly, realistic artificial worlds; (3) having developed the ability to create such worlds, there would be no absolute limit to the number of artificial universes that could be produced, and (4) that means, on the balance of probability, we are more likely to be living in one of the near-infinite simulated realities than base, physical reality.[18] That argument was later popularized and given a more detailed logical grounding by philosopher Nick Bostrom.[19]

The simulation hypothesis and its stricter philosophical formulation swiftly gained steam, becoming so significant that the famed American Museum of Natural History (AMNH) in New York City hosted a debate on the topic in its 2016 Isaac Asimov Memorial Debate. Participants in the debate included scientists from the Massachusetts Institute of Technology, University of Maryland, and Harvard, as well as a philosopher from New York University, and the event was hosted by Neil deGrasse Tyson, the famed director of the Hayden Planetarium and host of the rebooted *Cosmos*. The debate was so popular that tickets sold out in three minutes, and the YouTube video of it has been viewed more than 4 million times (as of October 2020). While none of the panelists asserted that we definitely are living in a simulation, merely engaging with the simulation hypothesis at the AMNH offers prestige to the hypothesis. To conclude, Tyson suggested that 'the likelihood may be very high' that we are living in a simulation.[20]

The importance of Apocalyptic AI is not limited to popular curiosity about the simulation hypothesis; it is a recurring theme in discussions of existential risk to humanity.[21] For example, Michio Kaku came to use the possibility of human extinction as a focal point for his writing. In *The Future of Humanity* (2018), he argues that humanity must emigrate to

---

[18] Moravec, 'Pigs in Cyberspace', 20–1; Moravec, *Robot*, 168, 173.
[19] Bostrom, 'Are We Living in a Computer Simulation?'
[20] American Museum of Natural History, '2016 Isaac Asimov Memorial Debate', 1 hour, 38 minutes, 55 seconds.
[21] There is a difference between existentialism as a philosophical endeavor and what early 21st-century advocates label 'existential risk' to humanity, which is a question of humanity's future existence rather than a question of the meaning of human existence.

space in order to ensure the survival of our species.[22] This marks a change in his worldview: in *Visions* (1997), he makes no note of mind uploading or other efforts to avoid extinction. As an important part of humanity's survival, Kaku detours into apocalyptic expectations of intelligent robots, mind uploading, and posthuman evolution.[23] Matching Kaku, Neal Stephenson pays his dues by (dismissively) mentioning both the simulation hypothesis and 'the Singularity Kool-Aid' in his novel *Seveneves*, which details humanity's effort to survive global catastrophe of the kind Kaku predicts.[24] One of the AMNH panelists, M.I.T. physicist Max Tegmark, became a leading voice in debating the role of AI in contemporary life by co-founding the Future of Life Institute to work towards positive developments of technology (i.e. to avoid catastrophes such as portrayed in dystopian science fiction).

The question of existential risk, powerfully intertwined with scientific futurism well before Tegmark, adds a distinct flavor to pop science that was not present in Moravec's watershed publications.[25] Tegmark's *New York Times* bestseller, *Life 3.0*, begins with the prospect of machine intelligence 'boot strapping' its way past human intelligence and declares the future of AI to be 'the most important conversation of our time'.[26] For Tegmark, the issue is whether superintelligent AI will be beneficial or catastrophic, and so he explores a panoply of possible futures—almost all emergent from or reacting to the Apocalyptic AI worldview.[27] For this analysis, Moravec and Kurzweil are both significant inspirations.[28]

In less dramatic fashion, Apocalyptic AI also influences conversations about virtual reality technologies. This is no surprise given Moravec's and Kurzweil's advocacy of virtual worlds, immortal virtual lives, and simulated resurrection. The previous chapter shows how these ideas intertwine in Rosedale's entrepreneurial work. Jim Blascovich and Jeremy Bailenson's

---

[22] Kaku, *The Future of Humanity*, 6.

[23] Kaku, *The Future of Humanity*, 126–9, 200–5, 218–20.

[24] Stephenson, *Seveneves*, 211.

[25] The popularization of existential risk emerged out of the growing awareness of the potential of AI, the dangers of climate change, the activity of groups such as the Lifeboat Foundation, and certain authors, such as Eliezer Yudkowski in the field of AI and Nick Bostrom in the broader transhumanist concern with threats to humanity.

[26] Tegmark, *Life 3.0*, 3–4, 22.

[27] Tegmark, *Life 3.0*, 161–248. Contrarily, Tegmark's concern that we might simply kill ourselves, such as through nuclear war, reflects concerns with no direct relation to Apocalyptic AI; see Tegmark, *Life 3.0*, 195–200.

[28] For example, see Tegmark, *Life 3.0*, 32, 155.

pop science introduction to virtual reality offers an obvious example of the iron horsemen's influence: the book's title is *Infinite Reality: Avatars, Eternal Life, New Worlds, and the Dawn of the Virtual Revolution*. While they do not wholeheartedly endorse the transferal of consciousness into machines, Blascovich and Bailenson refer to Kurzweil and do suggest that virtual simulation of a person might establish an immortal legacy for that person.[29] In their empirical research, Blascovich and Bailenson found that volunteers had a strong and persistent desire to see their avatars and personality profiles continue for centuries.[30] Even the suggestion of such a possibility clearly aligns with the longstanding human desire to attain immortality, and the influence of the iron horsemen is decisive in our cultural perception of how virtual reality technology unfolds.

Returning to Brockman's *Possible Minds*, we can identify a variety of ways in which Apocalyptic AI serves as backdrop for the speculations of scientists and engineers. The book includes essays by 25 eminent scientists, engineers, and others. The essays consider the state of AI today and the impact of Norbert Wiener's *The Human Use of Human Beings* (1950). Some contributors to the volume specifically reference the Singularity and many engage cosmic regime change—the evolution from humanity to superintelligent AI.[31] This variety of thinkers, tasked with discussing the contemporary relevance of Wiener's cautionary book, show that doing so happens with Apocalyptic AI almost always in the background and often in the foreground.

The scientific legitimacy granted to transhumanism and Apocalyptic AI by contemporary science produces powerful engagement in humanistic and theological endeavors also. In his wildly popular book *Sapiens*,[32]

---

[29] Blascovich and Bailenson, *Infinite Reality*, 141–2.

[30] Blascovich and Bailenson, *Infinite Reality*, 146; see also 260–1.

[31] On the Singularity, see Lloyd, 'Wrong, But More Relevant than Ever', 7–11; Dennett, 'What Can We Do?', 53; Ramakrishnan, 'Will Computers Become Our Overlords?', 183. On the transition to posthuman life, see Wilczek, 'The Unity of Intelligence', 74–5; Tegmark, 'Let's Aspire to More than Making Ourselves Obsolete', 79; Tallinn, 'Dissident Messages', 94; Deutsch, 'Beyond Reward and Punishment', 120; Griffiths, 'The Artificial Use of Human Beings', 132–3; Gershenfeld, 'Scaling', 169; Hillis, 'The First Machine Intelligences', 172–7; Gopnik, 'AIs Versus Four-Year-Olds', 230; Wolfram, 'Artificial Intelligence and the Future of Civilization', 284.

[32] The book's Wikipedia entry ('*Sapiens: A Brief History of Humankind*') indicates that as of 2017, *Sapiens* had been translated into 45 languages, was on *The New York Times* best-seller list, had received the National Library of China's Wenjin Book Award for 2014, was shortlisted for the 2015 book awards of the UK Royal Society of Biologists, and had been labeled one of *The*

historian Noah Yuval Harari concludes with a delve into transhumanist futurism, including a nod to Apocalyptic AI projections. Harari skims across research on cyborg implants, for example, and suggests that 'we cannot even grasp [the] philosophical, psychological, or political implications' of a future in which brain–machine interfaces become possible.[33] Harari then moves to discuss AI, mind uploading, and the Singularity.[34]

Well prior to Harari's work, Christian theological engagement was relatively swift in academic circles.[35] More recently, denominational Christian hierarchies have joined the conversation. For example, the Ethics and Religious Liberty Commission of the Southern Baptist Convention issued a statement of principles regarding evangelical Christianity and AI. Among other concerns, the document rejects the idea that machines can have dignity or personhood, the idea that AI might be a 'means of improving, changing, or completing human beings', and that AI might supplant humanity.[36]

Apocalyptic AI has left the fringe and entered mainstream pop culture. Streaming shows, such as *Black Mirror*, *Altered Carbon*, and *Upload*, bring these ideas to a wide audience, as do Hollywood movies such as *Transcendence* (2014) and *Her* (2013), which won an Academy Award for Best Screenplay. The widespread use of Apocalyptic AI in entertainment marks a radical departure from the 20th century, sharing and validating the unapologetic faith that the iron horsemen place in apocalyptic futurism.

Rapidly progressing technology and the prestige of the horsemen in science, business, and pop culture have confirmed the technological significance of Apocalyptic AI. Whether likely or not, the belief in cosmic transformation into the Mind Fire, virtual life, and uploaded salvation has become impossible to ignore. American scientists might have been dismissive of claims about such transcendent futures early in the 21st century,[37] but

Guardian's 'best brainy books of the decade'. The entry also notes that the book received mixed reviews from scholars.

[33] Harari, *Sapiens*, 407.
[34] Harari, *Sapiens*, 408–11.
[35] Noreen Herzfeld was a noted leader in this regard; see Herzfeld, 'Creating in Our Own Image;' Herzfeld, 'Cybernetic Immortality versus Christian Resurrection;' Herzfeld, *Technology and Religion*, 57–69.
[36] Ethics & Religious Liberty Commission, 'Artificial Intelligence: An Evangelical Statement of Principles'.
[37] For example, see Geraci, *Apocalyptic AI*, 46–7.

it has become impossible to reflect on our technological future without due consideration for the visionary claims of Moravec and his allies. Meanwhile, the chorus of believers such as those discussed in the last chapter has grown stronger in recent years.

Translations

The influence of the iron horsemen is undeniable in western cultural life, but little attention has been paid to how the ideas they champion have been transferred or translated into other cultural contexts. I've labeled Moravec, Warwick, Rosedale, and Kurzweil as iron horsemen in reflection of the Christian horsemen of the apocalypse: Death, War, Famine, and Conquest (who is sometimes depicted more nefariously as Pestilence). Those four are the harbingers of judgement and the onset of a new world in the Book of Revelation, and so I used them in the last chapter to reflect on four leaders in the apocalyptic worldview of artificial intelligence technology. For all their popularity in the U.S., early in the 21st century their influence was only just beginning to be felt in Bangalore's tech community. If AI technologies are to change the cosmos as the iron horsemen desire, they must first transform Earth: the religious visions of Apocalyptic AI must overcome or merge with their global competition.

My early fieldwork, conducted during five months in 2012–13, indicates an initial reluctance to embrace Apocalyptic AI in India's scientific communities, pop science publications, and popular literature.[38] Nevertheless, some Indians observe that technological change may be advancing at exponential rates.[39] In fact, out of four speakers at the 'Facets of AI' online workshop held by the National Institute of Advanced Studies in 2020, there were two speakers who mentioned the Singularity and the possibility of greater-than-human intelligence.[40] As we shall see in this chapter, more recent fieldwork, coupled with sustained attention to India's pop science and pop culture landscape, supports my prior claim that the mass consumption of smartphones would reshape the landscape

---

[38] See Geraci, *Temples of Modernity*, 131–64.

[39] For an early example of this, see Kumar, *Information Technology and Social Change*, 3.

[40] The event was held on July 15 and July 16, with two speakers on each day. The Singularity briefly featured in talks by Nithin Nagaraj on July 15 and LM Patnaik on July 16.

for transhumanism.[41] Early in the century, however, the minimal relevance of science fiction, the emphasis upon practical problem-solving,
and the dominance of mythological storytelling made the ground infer-
tile for transhumanism. As much of this has been discussed elsewhere,[42]
I limit myself simply to the limitations of science fiction and transhumanist speculation in 21st-century India. Once these limits are clear,
I will turn to the changes wrought by smartphones, Internet access, and
shifting trajectories in literary and popular science.

While India has a science fiction tradition that traces back more than
one hundred years, competition with mythology restricted the genre's
popularity among reading and film audiences until well into the 21st century. When asked about the relevance of science fiction, scientists and
engineers—even those individuals who were avid science fiction fans—
indicated that such stories are rather inconsequential in both science
and culture more generally. One roboticist explained to me that 'science
fiction has not really taken off in India ... maybe it is something cultural ... we already have a culture of storytelling that goes back a couple
of thousand years, right? So, I think of something like science fiction as a
separate idea; it, it cannot germinate in such a place where there is such
a rich tradition of meat'. A biologist reported that people have a 'richer'
connection to mythology than they do science, and another biologist—
when asked about the influence of science fiction on the scientific
community—replied 'hmm, mythology much more than science fiction,
would play a role in the sort of metaphors you think about, the way you
slice things up, and so on. So the first connection you make would be to,
some mythologies are a bit like science fiction ... those are the stories we
grow up with'. One final example: an engineer whose work crosses industrial design and robotics told me that 'popular science does not sell in
India ... science fiction, we don't read. I should take my comment back,
because we do read science fiction in the form of our mythology. Only
thing is we don't believe in it ... We don't think that way. I can think of our
mythology as our science fiction'. This perspective is so widespread that
even several years later it could be guessed at. In 2019, I asked an avowedly

---

[41] See Geraci, *Temples of Modernity*, 153–8.

[42] Geraci, *Temples of Modernity*, 131–58. Shiv Visvanathan briefly echoes my argument that
mythology has obstructed the growth of science fiction in India; see 'Forward', Kindle location 11.

transhumanist graduate student at the Indian Institute of Science to guess why scientists had previously shown so little interest in transhumanism. He replied: 'I think we have that in mythology', precisely echoing the scientists of my earlier research.

Without nurturing science fiction in science and in wider culture, futuristic speculation such as we see in transhumanism is nearly impossible. One Indian chemist avers that 'science is the graveyard of futurologists'.[43] A scientist risks a certain amount of notoriety for making futuristic promises, but there can be no doubt that—in reality—making futuristic promises is part and parcel with scientific trade. A great many scientists engage in such activity, and it frequently becomes a technique in the construction of cultural authority. Certainly that is the case for Apocalyptic AI thinkers.[44] In India, however, a biotechnology professor acknowledged in 2013 that transhumanism 'hasn't really ... penetrated into anything meaningful or substantial'. Curiously, all of this is true despite the fact that sometimes more accurate assessments of the digital future came from India than from the U.S.: in 1967, Lalit Kanodia of Tata Consulting Services predicted that the 'common man will be interacting with the computer in his daily life within the next decade or two'.[45] Compared to Thomas Watson's 1940s prediction that the entire world would have use for 'maybe five computers', Kanodia seems downright prescient. And yet, transhumanist futurism struggled to develop.

For two decades, D.K. Wadhawan, a fellow at the Bhabha Atomic Research Centre in Mumbai, was the lone prophet crying out the transhumanist dream in the wilderness of Indian science.[46] Wadhawan published two articles in the science education journal *Resonance* incorporating ideas from Moravec and Kurzweil, arguing in favour of an eventual transition to mechanical life. In addition, he peppers one textbook with references to superintelligent machines and uses the evolution of machine intelligence as the crux of a monograph on smart structures.[47] He concludes the latter by stating:

---

[43] Balaram, 'The Promise of Biology and Biotechnology', 427.
[44] Geraci, 'Cultural Prestige;' Geraci, *Apocalyptic AI*, 56–70.
[45] Bassett, *The Technological Indian*, 253.
[46] This allusion to prophets crying out in the wilderness references Biblical prophecy and is certainly not meant as a slur upon Indian science.
[47] Wadhawan, *Complexity Science*, 152, 160, 264 (all page numbers listed are for the PDF pages in the preprint version of the book); Wadhwawan, *Smart Structures*, 251–2, 254, 260–3.

Humans will probably coevolve with these 'artificial' superintelligences, via neural implants that will enable the humans to upload their carbon-based neural circuitry into the hardware they themselves were instrumental in developing. They will then live forever as bits of data flowing through 'artificial' hardware. That will mark a complete blurring of the distinction between the living and the nonliving.[48]

While the Oxford publication of *Smart Structures* can be seen as the most prestigious of his apocalyptic claims, its publication in the West makes the book less helpful for understanding the role of such thinking in India. And so an analysis of his Indian publications cannot be ignored.

In the first *Resonance* article, an essay on smart structures and materials, Wadhawan concludes with the claim that 'our machines (smart structures included) will evolve, gradually undermining the distinction between technology and nature, and between the living and the nonliving. The hardware and software will produce its own hardware and software, as needed and desired (by whom?!). What will be the role of human beings in such a world?'[49] The second essay, titled 'Robots of the Future', shows Wadhawan's commitment to Apocalyptic AI. After engaging contemporary strategies in robotics and computing (e.g. subsumption architecture, multicore processing, etc.), Wadhawan writes that if present trends continue 'by the year 2050 or so intelligent robots would have evolved to such an extent that they would take their further evolution into their hands'.[50] Following this, he mentions Moravec's mind uploading hypothesis and agrees with Moravec in claiming that 'it appears inevitable that, aided by human beings, an empire of *inorganic life* will evolve, just as biological or organic life has evolved. We are about to enter a *post-biological world*'.[51] Echoing Moravec's faith in a two-stage apocalypse, one where humanity will be financially at ease and relax into a 'comfortable tribalism', Wadhawan argues that 'intelligent robots will not only bring prosperity for all, but will also have highly salutary effects on the ecology of our planet'.[52]

---

[48] Wadhwawan, *Smart Structures*, 263.
[49] Wadhawan, 'Smart Structures and Materials', 41.
[50] Wadhawan, 'Robots of the Future', 75.
[51] Wadhawan, 'Robots of the Future', 77, emphasis original. Wadhawan's comments on mind uploading appear pp. 76–7.
[52] Moravec, *Robot*, 137. Wadhawan, 'Robots of the Future', 78.

From the onset of *Resonance* to 2019, Wadhawan's essays were the only two articles in the entire span of the journal that assess or acknowledge the claims of Apocalyptic AI, revealing them to be extreme outliers in the scientific community. Furthermore, they were themselves relegated to the academic and cultural fringe. According to GoogleScholar data, Wadhawan's 'Smart Structures and Materials' had been cited just 12 times by the end of 2018, and all the citations regarded the technical claims of the essay. Wadhawan's 'Robots of the Future' was cited just six times: that none of these citations came from Indians shows the trouble which transhumanism had in penetrating early 21st-century Indian intellectual circles.

Given the absence of transhumanist speculation in published and recognized areas, it should come as no surprise that scientists in India tended to dismiss apocalyptic visions of technology. During my initial fieldwork and a subsequent visit to the Indian Institute of Science in Bangalore in 2016, I was often informed—and in no uncertain terms—that the transhumanist project had no place in Indian science. A scientist in nanomaterials said of mind uploading that it 'might be popular in industry circles. But I think maybe with the common man and with the profound thinkers, soul probably is immortal'. A neurobiologist referred to speculation about mind uploading as 'nonsense' and a materials research scientist, upon hearing of my prior work on mind uploading immediately said 'oh! Like Ray Kurzweil!' and then asked me 'do many of us here believe in those things?' When I replied that the majority do not, he exclaimed 'oh, good!' and laughed heartily.[53]

Indian scientists' opposition to transhumanism may result from the direct influence of Hinduism. Adam Buben notes that despite the loud rhetoric of Christians opposing transhumanism, the radical longevity promised by the latter dovetails nicely with the panoply of health and life extension technologies already accepted by Christians (though he offers only minimal distinction between life extension and more radical transhumanist interests such as mind uploading).[54] Buben's main point,

---

[53] For a longer treatment, see Geraci, *Temples of Modernity*, 136–40.
[54] Buben, 'Personal Immortality in Transhumanism and Ancient Indian Philosophy', 71–2, 74–5.

however, is to indicate that he believes Hindu and Buddhist perspectives on human personhood are irretrievably opposed to transhumanist desiderata such as mind uploading: the transhumanist interest in prolonging the self prevents the self from realizing its true identity 'devoid of all these superficial and accidental particularities', an identity that participates in 'one unified whole' of reality.[55] So, while a variety of cultural and literary realities might slow the growth of transhumanism in India, it is also plausible that the metaphysical speculations of Indian religious belief could fundamentally disagree with transhumanism's focus upon personal, biological or post-biological immortality.

Although we see a great many texts in Hinduism that suggest one ought to seek release from the world, there are opposing traditions. In his *History of Hindu Chemistry*, for example, P.C. Ray mentions alchemical traditions that trace to the *Atharva Veda* and which propose strategies for prolonging life.[56] As in European alchemical practice, mercury is supposed as useful to 'make the body undecaying and immortal'.[57] Furthermore, a scientist whom I met in 2012 argued that the existence of the seven immortal *Chiranjivis* indicates that opposing immortality 'is not inevitable in the Indian belief system'.[58] So there are certainly countervailing tendencies in India despite the dominant stream in which bodily immortality conflicts with the paramount theological goal of *moksha* (release).

In its interrogation of technology, science fiction often puts to the test India's philosophical and religious beliefs about the soul. In his watershed science fiction novel, *The Calcutta Chromosome*, Amitav Ghosh retells the history of malaria research but does so in a quasi-magical world where a religious cult, using science to achieve transmigration of souls, guides the scientific process to its culmination. But such connections need not be one way. For Ghosh, traditional reincarnation enables medical progress, but the reverse is also possible. Anil Menon envisions a future where technology permits endless and deliberate reincarnation: individuals can medically become entirely new personalities and even individual selfhood exposes itself to technical intervention and transhumanist desire

---

[55] Buben, 'Personal Immortality in Transhumanism and Ancient Indian Philosophy', 76.
[56] Ray, *A History of Hindu Chemistry* vol. 1, 9.
[57] Ibid., 127. See also *A History of Hindu Chemistry* vol. 2, 29.
[58] I have previously noted this in *Temples of Modernity*, 140.

to choose the future.[59] As a final example, technology opens the door to divine incarnation in Gita Chandra's story, 'The Goddess Project'. Just as Indian mythology acknowledges the possibility of several gods' birth in earthly bodies, Chandra acknowledges there is no split between the technological triumphs of posthumanity and the divine, which comes to humanity incarnated in android bodies.

Contemporary Indian science fiction demonstrates an emergent potential for integrating religion and science towards transhumanist fulfillment; younger scientists and engineers, growing up in an era of such writing, occasionally embrace transhumanist ideas. Despite the reluctance to embrace transhumanism and the inconsistency in awareness of it among the older scientists, there were students and young tech workers who showed interest. A graduate student at IISc told me 'of course … many people inspired by that … we used to talk about a lot of science fiction or science fiction idea, a lot of futuristic idea. And we try to build. These kinds of futuristic idea basically … guide you, motivate you'. Other students agreed in part or in whole, often citing Hollywood influences. One might see these students as an iteration of William Gibson's oft-quoted maxim: 'the future is already here. It's just not very evenly distributed'.[60] As I've noted elsewhere, it was not just students, but young software engineers and others in the tech industry that showed a greater awareness of transhumanism and found some transhumanist promises compelling.[61] In 2012, the first inklings of transhumanism in Indian scientific culture were apparent even if they were not widespread.

In order to understand how transhumanism generally, and Apocalyptic AI in particular, will operate in Indian culture, we must first establish the ground rules for integrating such a vision. Apocalyptic AI is a perspective on cosmic transformation—the inauguration of a new world—and it emerged in a specific cultural context that focuses on linear views of history and singular moments of transformation. But as we saw in Chapter 1, Indians do not always see time in the same linear fashion as

---

[59] Menon, 'Shit Flower'.
[60] Gibson states this on National Public Radio, 'The Science in Science Fiction'. Although Gibson never expressed it in his writing, quoteinvestigator.com concludes that 'the evidence is strong that William Gibson used this expression' and the organization 'believes that he created it'.
[61] Geraci, *Temples of Modernity*, 137–9.

the Christians and post-Christians of the Euro-American West. Cosmic calendars are generally cyclic and perspectives on the world's evolution vary. So it is necessary to think about what cosmic transformation has looked like in India in order to think about its translation of Apocalyptic AI and the latter's perspective on human destiny. What cultural rules govern the move from a scientific community that largely ignored Wadhawan to one where Apocalyptic AI is taken seriously and engaged intellectually? While India's religious and philosophical ideas often diverge quite dramatically from those of the West, there is an intriguing similarity: as in Christianity's Book of Revelation, Indian transformation occurs when horses roam across the landscape.

Horses, not horsemen, and the significance of sacrifice

In Hindu religious texts, as with the biblical Revelation, horses inaugurate new kingdoms and carry conquest in their stride across the landscape. In India, however, the horse must be sacrificed, asphyxiated as part of a ritual's culmination. The *Ashvamedha* (horse sacrifice) held by kings in ancient and classical India reveals the important role that horses play in Hinduism and their significance for the kinds of world transformation dreamt of in transhumanist philosophy. Wendy Doniger notes that early in the *Rig Veda*, horses 'represented the "Aryas," as they called themselves' and in later scriptures a host of invading, conquering, or liberating peoples.[62] Early in the scriptural corpus of Indian religion, horses were already associated with repopulation, transformation, and social revolution. Much like the horsemen of Christian theology, horses directly occupy the space of cosmic transformation in addition to social and political change. For example, the mythological fiery mare at the bottom of the sea 'comes to symbolize the latent force of doomsday always poised to break out and destroy the universe.'[63] Symbolically opposite, Hinduism's cosmic savior Kalki, the tenth avatar of Vishnu, sometimes horse-headed but more often riding a white horse, will come to end *Kali Yuga* and restart the cosmic cycle at *Satya Yuga*. Horses

---

[62] Doniger, *On Hinduism*, 439, 446.
[63] Doniger, *On Hinduism*, 452.

are thus critical bearers of the Hindu apocalypse, but unlike the Christian formulations adopted by Apocalyptic AI advocates, Hinduism's horses cannot bring about transformation without sacrifice.[64]

Understanding the horse sacrifice is crucial to conceiving of transhumanism's translation to and reconfiguration within India, but this perspective is not without complications. On one hand, the *Ashvamedha* adds an obvious component of sacrifice to our considerations, one that might be lacking in western transhumanism's Christian heritage (where sacrifice is principally produced in the past by Jesus on the cross rather than in the present). For example, T. L. Vasvani, in a rather unusual journal of spirituality, engages the *Ashvamedha* as described in the *Brihadaranyaka Upanishad*, where he concludes by reflecting on the importance of service, of seeing the sacrifice as one that should transcend individual grasping and benefit all people.[65] A similar perspective could become a central contribution of Indian transhumanism. On the other hand, the ritual speaks also to a traditional Hindu rejection of immortality that we already noted in Adam Buben's response to transhumanism and Hinduism. The *Brihadaranyaka Upanishad* responds to the transhumanist goal of immortality, but of course confirms traditional Hindu aspirations towards release rather than unlimited life on earth. According to the text, he who fully understands the sacrifice 'conquers further death, death cannot overtake him.'[66] As Shankara avers in his commentary (8th/9th century CE), this means that 'after dying once he is not born to die any more.'[67] It may be difficult to sustain transhumanism's commitment to (unending) life extension for those who value extinction from a system of rebirth.

However, earthly immortality opposes most western modes of religious salvation also; transhumanists are already comfortable setting themselves apart from and in contrast to traditional religions. As such, Indians might well adopt key perspectives from their religious traditions without adherence to those traditions' stated goals (as has happened in the transition from Christian views of technology to transhumanist

---

[64] Even the mare in the sea is considered a sacrificial animal; see Doniger, *On Hinduism*, 457. In addition to being sacrificed, a horse often consumes that which is sacrificed; see Doniger, *On Hinduism*, 465, 466.

[65] Vasavani, 'Asvamedha Or the World Sacrifice', 55–6.

[66] Mādhavānanda, *Bṛhadāraṇyaka Upaniṣad* I.ii.7 (p. 36).

[67] Mādhavānanda, *Bṛhadāraṇyaka Upaniṣad*, 39.

views). In sum, the *Ashvamedha* offers little by way of immediate intellectual segue into transhumanism. We can look to it, however, as a clear example, perhaps the clearest example, of how Hindus have anticipated changing political and cosmic order to see the patterns that govern local worldviews.

In his analysis of the *Apastamba Srautasutra* (20.1–23), David Knipe succinctly describes the *Ashvamedha*.[68] The entire ritual process requires a year of time, during which the king is on a learning holiday and a horse roams the countryside, accompanied by warriors. Wherever the horse walks during his travels comes under the control of the king undertaking the ritual.[69] During the horse's absence, a series of ritual procedures take place, focusing on learning but also on the integration of important members of the community into the kingdom whose boundaries get defined by the ritual.[70] Upon the horse's return, the final and most important part of the ritual takes place.[71] Hundreds of animals are sacrificed. The horse is washed and adorned with gold, silver, and pearls by the kings' wives, ghee is put in its mane and tail, and it is attached to a chariot. The horse is then smothered with a blanket while hymns are sung. The queens circumambulate the horse while the chief queen supposedly lies with her legs entwined around the horse's phallus or even engages in intercourse with the dead horse.[72] The horse is butchered, and the king sits on his throne. A denouement of concluding rituals follows this dramatic high point.

[68] Knipe, *Vedic Voices*, 234–7. For another excellent summary of the various ritual steps, see Law, 'The Horse-Sacrifice and Its Political Significance'.

[69] Doniger notes that the soldiers might be said to goad rather than follow the horse on the king's quest to claim land (*On Hinduism*, 439). There exist arguments on whether only 'paramount sovereigns' can complete the *Ashvamedha*, as occasionally noted in scripture. The fact that a local ruler might arrest the horse and thereby challenge the kingship, as noted by Sur in 'Asvamedha—A Rejoinder', indicates that the ritual seems more likely a way for a ruler to *justify* his claim to being a paramount sovereign rather than a ritual practiced on account of that fact.

[70] Karmakar, 'The Pariplava (Revolving Cycle of Legends) at the Asvamedha'.

[71] On the importance of the final days, see Law, 'The Horse-Sacrifice and Its Political Significance', 634.

[72] In his otherwise copious detailing of the *Ashvamedha*, Law finds the whole matter so distasteful that he asserts 'decorum does not permit me to give here its details' ('The Horse-Sacrifice and Its Political Significance', 639). Ramachandran, on the other hand, swiftly notes that 'the crowned queen lies down by the side of the dead horse … and she unites with it' ('Aśvameda Site Near Kalsi', 11). Neither author indicates any doubt as to the likelihood of the queen willingly copulating with a dead horse or of the king willingly having his favourite wife do so. Despite the obvious reasons one might doubt that this truly happened, Jamison declares that 'the texts leave no doubt as to what physically she is supposed to do, and it is not merely symbolic' (*Sacrificed Wife/Sacrificer's Wife*, 65). I suspect that Jamison and others have fallen into an intellectual trap common to academics: credulity. In his seminal essay 'Archeology and Protestant Presuppositions in the Study of Indian Buddhism', Gregory Schopen argues that scholars of

The *Ashvamedha* sacrifice has not been held in approximately 200 years, not since Savai Jayasingh in the early 18th century,[73] but in theological and ritual terms, it remains an interesting standard for considering Hindu political and cultural transformations. Naturally, while no one sought to drive the British away with a horse sacrifice, nationalists referenced the *Ashvamedha* as part of Indian identity formation: V.D. Savarkar, inventor of the concept of *Hindutva*, notes the return of the 'Horse of Victory' that marked the successful unification of India's land and, importantly, peoples.[74]

In ancient Hindu texts, the king is often aligned with cosmic order (such as in the *Rajasuya* consecration ritual[75]), and the king himself is perceived as permeating the cosmos through ritual practices.[76] And so the transformation of kingship through coronation or through expanded rulership involves necessary transformations of the cosmos. According to some ancient authorities, from the *Mahabharata* to political inscriptions, this relationship means that a sufficiently righteous king overcomes *Kali Yuga*, establishing a righteous cosmos.[77] The ritual ought to produce both political and spiritual gains, such as the acquisition of power and the removal of sins.[78] It may even have had an 'elevatory' nature, raising the station of outsiders or low-status insiders.[79] The capacity to transform

Buddhism—and here, I extrapolate to the wider academic community—believe what is written even in the face of common sense or absolutely contrary archeological evidence. J.Z. Smith similarly demonstrates that scholars of ritual are susceptible to believing outright and often absurd falsehoods which their research subjects claim to be true (*Imagining Religion*, 59–61). Ultimately, there are simply too many reasons why neither the queen nor the king would desire actual copulation between the horse and queen to believe that anyone actually felt compelled to take the text literally and engage in bestiality.

[73] Knipe, *Vedic Voices*, 234. Ghosh has the name as Sowae Jaya Simha in 'Aśvameda and Rājasūya', 763. Although the *Ashvamedha* ritual is long out of practice, Koppers attempts to tie the ritual to a practice still current during the time of his research in 'The Mundas and the Sidoli Feast of the Korkoos', 209–11.

[74] Savarkar, *Hindutva*, 11–2.

[75] See Heesterman, *The Ancient Indian Royal Consecration*, 205. This ritual also involves a traveling horse and at least one scholar believes it to be more significant than the *Ashvamedha* (Ghosh, 'Aśvameda and Rājasūya').

[76] Heesterman, *The Ancient Indian Royal Consecration*, 193. Law notes that the *Vajapeya* ritual, in which the king maintains a connection to horses by riding in a victorious chariot, indicates the king's identity with Prajapati, lord of creatures; 'Some Vedic Rituals and Their Political Significance', 534. Ramachandran notes the same identity as well as a more general comparison to the gods in the *Ashvamedha* in 'Aśvameda Site Near Kalsi', 10, 14.

[77] See González-Reimann, *The Mahābhārata and the Yugas*, 118–9, 129–32.

[78] Law, 'The Horse-Sacrifice and Its Political Significance', 634. See also Ramachandran, 'Aśvameda Site Near Kalsi', 12.

[79] Ramachandran, 'Aśvameda Site Near Kalsi', 30.

outsiders to insiders or to purify the sinful offers some key as to the cognitive moves necessary as transhumanist thinking encounters the logic of Hindu ritual.

More even than the king's alignment with the cosmos, the sacrificed horse is itself representative of the cosmos; the sacrifice is doubly concerned with the state of the world. This is quite clear in the *Brihadaranyaka Upanishad*, which begins 'Om. The head of the sacrificial horse is the dawn, its eye the sun, its vital force the air, its open mouth the fire called Vaiśvānara, and the body of the sacrificial horse is the year. Its back is heaven, its belly the sky, its hoof the earth, its sides the four quarters, its ribs the intermediate quarters, its members the seasons, its joints the months and fortnights', etc.[80] Thus, the horse itself encompasses all of space and time; its sacrifice is a transformation and a restoration.

Unless challenged, the path of the *Ashvamedha* horse inaugurates dominion and transforms land from one political domain to another; at the same time, the horse can bring about a spiritual transformation. As noted, the *Ashvamedha* can absolve humanity of sinfulness. Perhaps this is why Yudhisthira hosts an *Asvamedha* in the *Mahabharata*. Having already settled his kingship, Yudhisthira clearly has no use for the ritual's political significance. But filled with guilt over warfare and observing that the world needs a new beginning, Yudhisthira hosts the sacrifice as a way of overcoming anomie and sin.[81]

Although horses are rare in India, they remain vitally tied to many religious practices and continue to mediate relationships of indigeneity and foreignness. Noting that clay horses—said to 'be ridden by spirit riders who patrol the border of villages'—are commonly found across India, Doniger notes that 'the villagers do not express any explicit awareness of the association of the horses with foreigners; they think of the horses as their own'.[82] Critically, this usage makes those horses uniquely Indian. While horses may have been imported and their presence representative of conquerors or liberators, the meaning of the horses in the village is, of course, local and locally relevant. So the horse is simultaneously local and foreign, conqueror and liberator, bearer of kingship, and the sacrifice that

---

[80] Mādhavānanda, *Bṛhadāraṇyaka Upaniṣad* I.i.1 (p. 8).
[81] Jamison, *Sacrificed Wife/Sacrificer's Wife*, 76–7.
[82] Doniger, *On Hinduism*, 449.

inaugurates a new era. Could a worthier metaphor for the apocalyptic visions of artificial intelligence be found?

Strangely—considering that transhumanism is just beginning to be articulated and reshaped in India—Hinduism has already served as the bearer of spiritual transhumanism outside of India. In her richly illustrative study of Cosmism, transhumanism, and immortalism in 21st-century Russia, Anya Bernstein notes how Dmitri Itskov's 2045 Avatar Project, which aims towards mind uploading, was nurtured by a Hindu ashram. Itskov 'was a devoted disciple and wanted to help [the ashram's guru, Swami Vishnudevananda] realize his idea of immortality through science'.[83] Clearly, we must reject preconceived notions that transhumanism flows in straightforward paths from West to East! Rather, complex networks of religion and science allow ideas about science and technology to circulate among people who reconstitute them according to local interests.

Just as Indians adopted and reconfigured the horse, and even as Hinduism plays a role in cultural construction of transhumanist aspirations outside India, the same must be expected of AI and Apocalyptic AI in India. What enters India as foreign shall be put to local service. Sundar Sarukkai argues that people use intersections with religion to enlarge 'the domain of beliefs we hold about technology in order to enable us to deal with it in a manner suited to us'.[84] Perhaps this sheds light on the moment when a student of human–computer interaction asked me 'What is god right now? So it's a very subjectival thing … About Kalki, the day of Kalki: so my suspicion is AI is Kalki. I mean a truly artificial intelligent being is Kalki, maybe'.[85] Jayan Prasad, who operates Singularity Café in Chennai, agrees that such a belief could become popular: 'right now', he told me, people believe Vishnu 'is going to return as Kalki and many people would say AI is Kalki'.[86]

---

[83] Bernstein, *The Future of Immortality*, 53. For more on Itskov's project and its religious connections, see Asprem, 'The Magus of Silicon Valley', 407–8.

[84] Sarukkai, 'Culture of Technology and ICTs', 47, emphasis removed. This process is complex, perhaps especially in the domain of AI. After all, even the appearance of a robot must be aligned with cultural expectations (see Złotkowski, Khalil, and Abdallah, 'One Robot Doesn't Fit All'). See also Arnold, *Everyday Technology*, 6.

[85] Comment during question-and-answer period of Geraci, 'Religion, Technology, and Human-Computer Interaction', 1:22:38. Intriguingly, Antoinette DeNapoli describes how North Indian *sadhus* often describe human–technology interactions as Kalki, with the expectation that our merger with technology opens the door for a redemptive future; see 'Dharm' is Technology'.

[86] J. Prasad, email correspondence with the author, 26 December 2019.

From Daedalus to Kalki

The cultural conditions for a student to propose that Kalki might arrive as AI emerge from a growing awareness of transhumanist themes in India and a shift in the social landscape. It would come as a surprise to many scientists, perhaps even Wadhawan, but Indian authors have flirted with transhumanism since at least the early 1930s. But those tentative early steps disappeared into history, with early 21st-century Indians unaware of them. That heritage and the work of early advocates such as Wadhawan position Indians to get involved in global debates about futurism, transhumanism, and AI. Heading into the third decade of the 21st century, one finds Indian versions of transhumanism appearing in pop science, science fiction, and communities that openly combine religion and science.

Appreciating India's growing interest in transhumanist promises requires that we look first to the technological futurism of J.B.S. Haldane, a British scientist who spent his final years in India. Although primarily famous for his work in population genetics, the mathematics of biology, and for tireless labours popularizing science, Haldane is also known for a lecture delivered to the Heretics at Cambridge and subsequently published.[87] This book, *Daedalus, or Science and the Future*, is among the principal building blocks of 20th-century transhumanism, and its impressive accuracy casts a rosy glow on the entire enterprise of futuristic prognostication. The book was a stunning success, selling 15,000 copies in its first year and going through 10 printings between its initial publication in 1923 and 1926.[88] Haldane's futuristic expectations for human evolution date back to his childhood education in Eton, where he wrote in his diary that scientific progress means that 'Man ... ought to be a god, or something very like it, if he exists 10^6 years hence'.[89]

---

[87] Curiously, not long after Haldane's Heretic Society speech his father was invited to deliver the Gifford Lectures on science and theology, and these also caught some attention in India; see Tattvabhushan, 'Dr. Haldane on the Immortality of the Individual'.

[88] See Dronamraju, 'Introduction', 1.

[89] Quoted in Subramanian, *A Dominant Character*, 76. Haldane may be influenced by Élie Metchnikoff, who wrote about alleviating the 'illness' that is old age in *The Nature of Man*. In an email conversation with me (7 March 2020), Subramanian reports that Haldane wrote about indefinitely expanded lifespans in November of 1910 and of Metchnikoff's book in April 1911. However, Subramanian notes two further complications: (1) Haldane may have purchased the Metchnikoff book as much as six months prior and (2) Haldane's discussion of extended lifespans happens in a diary entry that is 'sort of a summation of his life and his person so it's possible he had the idea before November 1910'. Additionally, Metchnikoff makes no claims about

In *Daedalus*, Haldane predicts ubiquitous urban electricity, near instantaneous global communication networks, wind and solar power (and the use of batteries to make them viable), neuropharmacology, lab-grown foods, attempts at interplanetary communication, genetic engineering, in vitro fertilization, and artificial wombs.[90] Haldane made further claims regarding the enhancement of the human species by genetic manipulation and the eradication of disease, and vouches for 'man's gradual conquest ... of his own body' though he does not promise an end to death.[91] These transformations, Haldane predicts, will necessitate that scientists take charge not only of the body but of the soul: science must kill religion and refashion morality so that humanity can form a global community.[92]

Haldane's publishers also brought forth a contemporary lecture by the Indian philosopher Sarvepalli Radhakrishnan, and this book begins India's halting engagement with transhumanism. In *Kalki, or The Future of Civilization*, Radhakrishnan seemingly offers a riposte to Haldane, arguing that science is certainly bringing about social transformation but that religion is the key to navigating our shared future.[93] Radhakrishnan notes that civilization itself is in a process of transition and recognizes that science is 'one of the chief factors' in this; moreover—preceding Kurzweil and others by decades—he notes of science that 'its pace of progress has

indefinite lifespans or posthuman evolutionary conditions. Metchnikoff's position is really that we could eliminate many of the ailments of old age and ensure that we live comfortably until we are ready to die (*The Nature of Man*, 262–84, 288–302). And so we may be looking at curious intersections rather than historical causes. A similar intersection is worth noting in Haldane's friendship with George Bernard Shaw, whose five play sequence *Back to Methuselah* of 1918–20 ends with predictions of a future posthumanity living immensely long lives as bodiless intelligence. Haldane's friendship with Shaw is noted in Mahanti, 'John Burdon Sanderson Haldane'. I am not aware of any effort to trace the philosophical influences between the two.

[90] Haldane, *Daedalus*, 18, 19, 23, 24, 35–7, 38, 40, 59, 63–4, 69, 70–1. I should note that Haldane makes other, less auspicious, claims, such as that eugenics will be a productive aid to society (p. 57) and that science will judge the merits of spiritualist claims—though he does not suggest that proof of ESP, etc. is likely (p. 76).

[91] Haldane, *Daedalus*, 70–1, 73, 82. Much later, Haldane agreed that immortality might be possible, but he did not see this as possible within the next ten thousand years; see 'Biological Possibilities In the Next Ten Thousand Years', 342.

[92] Haldane, *Daedalus*, 90–3. Again, Haldane may be influenced by Metchnikoff, who debates whether science can replace religious morality in *The Nature of Man*, 218–27, 285–9, 302. Evidence that Haldane read Metchnikoff can be seen in the imbroglio he caused by sharing the book with his classmates at Eton; see Subramanian, *A Dominant Character*, 77.

[93] *Daedalus* and *Kalki* share the same structure in the title, which they take from the To-day and To-morrow book series published by Kegan Paul, Trench, and Trubner. Before I found Radhakrishnan's book, I had titled this section 'From Daedalus to Kalki;' my discovery of Radhakrishnan's work put a delightful stamp of authenticity to the title.

become latterly too fast and its range too wide and deep for our quick adaptation'.[94] Like Haldane, Radhakrishnan acknowledges the force of biological explanations, works within an evolutionary framework, and expects that humanity will develop beyond: 'Our arrival on this planet is of recent date. No wonder we are only half-civilized. There is plenty of time ahead of us ... If we go on progressing, not only physically and mechanically but also mentally and spiritually, the prospect for humanity is great indeed. I am optimist enough to hope that the present upheaval will in the end promote the good of the world'.[95] Interestingly, Radhakrishnan and Haldane would become correspondents upon the latter's arrival in India.[96]

Thanks to his evolutionary stance, Radhakrishnan takes a positive view of humanity's opportunity for progress; in this, he sounds almost exactly like Moravec, Kurzweil, and their colleagues. He writes that

Man is neither the slave of circumstances nor the blind sport of the gods. The impulse to perfection working in the universe has become self-conscious in him. Progress *happened* in the sub-human world: it is *willed* in the human. Conscious purpose takes the place of unconscious variations. Man alone has the unrest consequent on the conflict between what he is and what he can be. He is distinguished from other creatures by his seeking after a rule of life, a principle of progress.[97]

Conceptually, the transition from blind nature to human choice is at the root of 20th and 21st-century transhumanism. Although Radhakrishnan is not a transhumanist, nor does he prefigure them, he shares with them a sense of human destiny and control.

[94] Radhakrishnan, *Kalki*, 7, 7–8. It is unclear whether Radkhakrishnan viewed his work as a response to Haldane; in his biography of Radhakrishan, Gopal notes that Radhakrishnan was invited to write the book but does not indicate any motive on the part of the publishers (*Radhakrishnan*, 88–9). It is perhaps telling, however, that Radhakrishnan had previous interactions with Haldane's father (and thus possibly Jack Haldane as well); see *Radhakrishnan*, 41, 69, 73, 76, 79.

[95] Radhakrishnan, *Kalki*, 53.

[96] Subramanian, *A Dominant Character*, 312. In an email to me (6 March 2020), Subramanian states that he has no sign of correspondence between the two at the time of *Daedalus* or *Kalki* and that every correspondence found is from Haldane's time in India. Further, Subramanian checked with Radhakrishnan's grandson, who agrees that this acquaintance did not begin until the 1950s. I am indebted to Mr. Subramanian for his insight and help.

[97] Radhakrishnan, *Kalki*, 63, emphasis original.

But for Radhakrishnan, human destiny hinges upon spiritual rather than scientific progress. He shows indifference to Haldane's transhumanist desiderata, as when he suggests that 'suffering is not an accidental accompaniment of life, but is central to it'.[98] He acknowledges the transhumanist goal of ascending to a higher evolution, but for Radhakrishnan, evolution leaves us human even as it moves us away from our basest nature. 'Every individual soul is an undeveloped entity, which, though not wholly emancipated from the animal instincts, is yet capable of transmuting them. By a willing acceptance of the commanding claims of spirit and a discipline of our nature into conformity with its law, we achieve growth'.[99]

Radhakrishnan declares that 'a community which is almost entirely preoccupied with life and body, physical and economic existence, scientific and technical efficiency to the exclusion of the higher humanistic ideals of mind and spirit is not truly civilized. Body, mind, and spirit form distinguishable aspects of an inseparable unity. Human nature is all of a piece, and unification of the three is the true aim of civilization'.[100] Radhakrishnan, speaking regularly of the distinction between animals and humankind, pursues a future in which we become more fully human, rather than posthuman. He concludes his book in the precise opposite of Haldane:

Religious idealism seems to be the most hopeful political instrument for peace which the world has ever seen. We cannot reconcile men's conflicting interests and hopes so long as we take our stand on duties and rights. Treaties and diplomatic understandings may restrain passion but they do not remove fear. The world must be imbued with a love of humanity. We want religious heroes who will not wait for the transformation of the whole world but assert with their lives, if necessary, the truth of the conviction 'on earth one family', heroes who will accept the motto of the great Stadtholder: 'I have no need to hope in order to undertake; nor to succeed in order to persevere'.[101]

---

[98] Radhakrishnan, *Kalki*, 61. I do not suggest that Haldane believed in a future utopia without suffering. In fact, he indicates clearly that new travails are to come; see *Daedalus*, 87.
[99] Radhakrishnan, *Kalki*, 62.
[100] Radhakrishnan, *Kalki*, 42.
[101] Radhakrishnan, *Kalki*, 96.

For Haldane, science must destroy religion and offer a path towards moral improvement and technological progress. For Radhakrishnan, a pluralistic and welcoming religious quest for our individual and composite betterment offers us a chance to evolve beyond our present state.[102]

The earliest engagement with what would be called transhumanism comes from an Indian review of Haldane's *Daedalus*. In an essay published by *Prabuddha Bharat* (the publication arm of the Ramakrishna Order begun by Vivekananda), Swami Ashokananda positions Haldane's optimistic gloss on science against a pessimistic essay penned by Bertrand Russell. For Ashokananda, however, the transhumanist promises are unimportant compared to the question of whether science endangers religion. Arguing that it does not, he claims that any scientific assault on religious obscurantism is a good thing and that religion will persist into the future.[103] But though he mentions Haldane's futuristic speculations, he offers no gloss upon them and even suggests that Haldane does 'not expect that there will be any of [his] readers who will believe [his] prophecies to be literally true'.[104] Instead, Ashokananda concludes that 'the Divine will awaken' and that 'people have grown tired of materialism and want something beyond the bread and butter interests of life'.[105] Strangely, the tendency to transform conversations about futuristic technology to other subjects persisted into the 21st century.[106] In this case, Ashokonanda reads futuristic promises and redirects the issues towards spiritual salvation.

Contemporary to Ashokananda, a British-born but American theologian publishing in India's *Modern Review* observes that some scientists believe it will be possible to significantly, perhaps radically, increase human lifespans but discards such directions in favour of spiritual growth.[107] For him, growing old is not so much a matter of the body but a process to be avoided in the mind and spirit. We see, therefore,

---

[102] Perhaps because Radhakrishnan pursues a pluralistic view of religion, he never actually mentions Kalki, the avatar of Vishnu, in his book. Not once does the name appear outside of the essay's title.

[103] Ashokananda, 'Science and the Future and the Future of Science', 319–20, 322–3. Ashokananda is certainly not unique among Indians in believing that science can purify religion; see Maitra, 'The Hope of Immortality', 597–8.

[104] Ashokananda, 'Science and the Future and the Future of Science', 316, 317.

[105] Ashokananda, 'Science and the Future and the Future of Science', 323.

[106] See Geraci, *Temples of Modernity*, 136–7.

[107] Sunderland, 'Why Need We Ever Grow Old?'

a complex intertwining of scientific futurism with religious and cultural conservatism operating on the fringes of Indian intellectual life.

Haldane was an important scientist and popularizer of science in India, but no horseman of the apocalypse there. K.R. Dronamraju, introducing Haldane's essays in *What I Require from Life*, writes of the time that:

> Haldane's essays were read widely in India. He rapidly became well known to the lay public. Photographs and cartoons of Haldane in his Indian attire were frequently seen in the Indian newspapers, and he used to receive a lot of mail from readers, which he answered promptly. He enjoyed the pantheon of Indian religions and philosophies. It is quite correct to say that Haldane enjoyed India and India enjoyed Haldane.[108]

During his time in India, Haldane spent years writing pop science articles and in these one can see occasional hints of transhumanist futurism.[109] Apart from offhand references to breeding new features into humanity, however, Haldane shows little fervor for grand speculations or evolutionary transformation. Nevertheless, his participation in a 1963 Ciba Foundation conference on 'Man and His Future' and the publication of his presentation 'Evolutionary Possibilities In the Next Ten Thousand Years' show that Haldane retained a clear interest in futurism and human destiny: the essay mentions human cloning, interplanetary colonization, genetic engineering, and technological intervention in human evolution.[110]

While he lived in India, Haldane's scientific colleagues largely avoided discussing futuristic speculation. Dronamraju explained to me that he was aware of *Daedalus* when Haldane arrived in India, but he glanced at it only 'briefly. That may sound strange today, but my attention was focused on other immediate subjects which required my attention promptly'.[111] Similarly, he reports that 'scientists in India were too conservative to discuss speculations. They were afraid it might harm their career in science'.[112] Dronamraju was a student and then colleague of Haldane before continuing his professional career in the U.S.; long after Haldane's

---

[108] Dronamraju, *What I Require from Life*, 128.
[109] For example, Haldane, 'Some Reflections on Non-Violence', 139.
[110] Haldane, 'Evolutionary Possibilities In the Next Ten Thousand Years'.
[111] Dronamraju, email correspondence with the author, 1 June 2019.
[112] Dronamraju, email correspondence with the author, 1 June 2019.

death, he compiled and edited Haldane's work into several volumes, including the one mentioned earlier. While it may have taken time to incubate, Dronamraju believes that Haldane's impact was seminal: 'it was Haldane who started Indian scientists to think more boldly and speculatively ... I think it has a lasting influence'.[113]

The extent to which Haldane changed (or didn't) the course of Indian science is hard to uncover at this late date; but without question, his impact was far greater in the realm of science popularization than science futurism. It is telling, for example, that in his wonderful biography of Haldane's science and politics, Samanth Subramanian makes no mention of transhumanist futurism during Haldane's years of life in India.[114] Given Haldane's approach in India, it is no surprise that Indian coverage of him rarely engages with transhumanism or futuristic speculation.

Mid-20th-century rationalist communities in India knew of Haldane and interacted with him, but they did so without regard for his futurist speculation. Upon the launch of *The Indian Rationalist* in 1952, for example, Haldane wrote a letter of greeting (presumably solicited) which indicates his awareness of their efforts, but in it he mentions nothing about futurism.[115] The following volumes of the journal show symmetry: when Haldane moved to India, the editor wrote a welcome letter to him which made no mention of his futurism, and in the journal's final year, Haldane published a message which, once again, made no conjectures about the evolutionary future of humankind.[116]

Even into the 21st century, one finds references to Haldane that do not engage his predictions. For example, in the pop science magazine *Dream 2047*, the very title of which seems to indicate an interest in futurist speculation, a fairly detailed biographical essay on Haldane completely omits reference to *Daedalus* or Haldane's futurism.[117] Haldane certainly did not launch a popular transhumanist crusade and evidently made little effort to develop transhumanism even though his friend and colleague, Julian Huxley, was doing exactly that in Europe.

---

[113] Dronamraju, email correspondence with the author, 1 June 2019.
[114] Subramanian, *A Dominant Character*, 284–331.
[115] Haldane, untitled greetings, 5.
[116] Ramanathan, 'Welcome Haldane;' Haldane, 'Message from J.B.S. Haldane'.
[117] Mahanti, 'John Burdon Sanderson Haldane'.

Influenced by Haldane, both Julian and his brother Aldous Huxley engaged with futurist visions of humanity. By World War I, at least, the Huxleys were conversant with Haldane's ideas, and by the 1920s, Huxley was also predicting technological advances that would reshape our species.[118] Interestingly, Aldous—at least—was also reading the work of Nikolai Berdyaev around this time.[119] Berdyaev is among the Russian exiles deeply influenced by Nikolai Fedorov, the principal philosopher behind Russian Cosmism. Berdyaev rejected Fedorov's focus on the past in favour of a creative future but joined Fedorov in looking for humanity to overcome death.[120] Whether Julian Huxley was in this way or another connected to Cosmism is unclear, but it seems plausible that he would have been at least moderately familiar with Berdyaev, at least.

Just as Haldane found an audience among Indians, so too did Huxley (though more modestly). Huxley coined the term transhumanism in its modern sense, but he differed from Haldane in seeing this movement as essentially religious: 'What the world needs is an essentially religious idea-system, unitary ... charged with the total dynamic of knowledge old and new, objective and subjective, of experience scientific and spiritual'.[121] Clearly drawing on Auguste Comte's Religion of Man, Huxley even believed that transhumanism could have rituals, temples, a priesthood, and other elements typical of a religious community.[122] Humanity, according to Huxley, has a 'cosmic office' and 'can, if it wishes, transcend itself—not just sporadically, an individual here in one way, an individual there in another way, but in its entirety, as humanity. We need a name for this new belief. Perhaps transhumanism will serve: man remaining man, but transcending himself, by realizing new possibilities of and for his human nature'.[123]

Huxley visited India and was a person of great interest to India's rationalist communities, but those communities evidently ignored or missed his interest in the future evolution of humanity. After Huxley's tour in the early 1950s, rationalists attempted to gain access more

[118] Deese, We Are Amphibians, 2–3, 14.
[119] Ibid., 62.
[120] Young, The Russian Cosmists, 138.
[121] Huxley, Religion without Revelation, unpaginated preface.
[122] Huxley, Religion without Revelation, 209.
[123] Huxley, New Bottles for New Wine, 14, 17.

intimate than attending lectures but found themselves stymied in this.[124] S. Ramanathan, the editor of *The Indian Rationalist*, reported on Huxley's open-minded approach to scientific study of extrasensory powers and other arcana but provides no insight into Huxley's views about the future of humankind.[125] The rationalist communities did have opportunity to engage with Huxley's transhumanism (or 'evolutionary humanism'), but they chose not to do so. Huxley was clearly aware of the journal, for he wrote a brief letter to the editor in its first year, a letter which was included in an editorial note.[126]

The editor and contributors to *The Indian Rationalist*, however, found themselves deeply concerned with Huxley's favourable use of the term 'religion' and criticized that without investigating his sense of human evolution.[127] The journal even reprinted one of Huxley's essays in which he advocates for a unification of science and religion as 'scientific humanism' in which 'better fulfillment would be the key idea, and man would appear as the instrument of destiny on this planet, the agent through whom new possibilities can be realized for the mysterious cosmic process'.[128] He concludes with the expectation that his grandchildren will discuss a 'religion based on evolutionary humanism'.[129] Even with direct opportunities for interrogating Huxley's ideas (admittedly expressed more vaguely than in many of his other writings), the rationalist community responded with only one letter to the editor: a response that quibbles with his terminology.[130] Similarly, the readership of *The Indian Rationalist* had nothing to say on the publication of a resolution espoused by the International Humanist Congress—and clearly derivative from the thoughts of its president, Huxley—stating that 'liberated from fear, the energies of man will be available for a self-realization to which it is impossible to foresee the limits'.[131]

---

[124] Ramanathan, 'Julian Huxley', 53. Although this article—like many in *The Indian Rationalist*—has no author attributed, one can reasonably expect that the editor of the journal is the author.

[125] Ramanathan, 'Julian Huxley', 54.

[126] Ramanathan, 'Humanism and Rationalism', 41.

[127] Ramanathan, 'Humanism and Rationalism', 42; Smelters, 'Materialism', 71; Isenberg, 'Call it Rationalism', 89. It may be worth noting that the responses from Smelters and Isenberg are from an Australian and an American, not Indians.

[128] Huxley, 'Morality without Religion', 115.

[129] Huxley, 'Morality without Religion', 115.

[130] Smelters, 'Morals without Gods', 140.

[131] International Humanist Congress, 'The Humanist Appeal', 48.

Despite these halting engagements with Haldane and Huxley in the 20th century, we can witness the 21st-century rise of technological futurism in the scientific and literary environments of India. Technological shifts enable new cultural perspectives, which of course return to affect technology. In the case of India, both pop science and science fiction have gained increasing prominence in the opening decades of the 21st century, and both increasingly engage with futurist speculation in technology. The improving prospects for Indian science fiction are visible in the dramatic popularity of Shankar's film *2.0*, the 2018 sequel to the 2010 film, *Enthiran*. While the latter was a commercial success, the former was meteoric. It set a record as the most expensive Indian film ever made and—as of 2019—was the second highest grossing Indian film ever made.[132] The changing fortunes of science fiction ensure that transhumanist futurism has an audience in India.

After a brief lapse in interest after Ghosh's *Calcutta Chromosome*, Indian authors began earnestly considering transhumanist claims around the same time that the Internet became more readily available there. Anil Menon's *The Beast with Nine Billion Feet* (2009), for example, describes the possible social outcomes of genetic engineering, but it may be important to remember that Menon was living in the U.S. at the time of publication. Within India, Samit Basu's *Turbulence* (2010) and *Resistance* (2014) also engage the possibility of transhumanism driving political and economic conflict. Other authors cleave to themes made popular by transhumanist advocates: short story examples include videogame avatars that mimic their creators' personalities when people go offline, an AI that becomes god while human beings are tied cybernetically together in a wireless communication network, and a political struggle over android rights.[133]

Importantly, Indian authors have drawn on local culture and religion in order to contemplate the nature of AI and related digital technologies. For example, in a short story published in *Breaking the Bow* (edited by Menon and Vandana Singh), one author uses digital technology to reconsider the classic epic of the *Ramayana*. In the story, the demon king

---

[132] Wikipedia, '*2.0*'.
[133] Baliga, 'The Collision of Parallels', 325; Chawla, 'The God Link', 353; Satyamoorthy, 'Dawn', Kindle location 151.

Ravana mixes 'new science with old-fashioned magic' and prayer to the gods to simulate a near-infinite number of worlds so that he can win Sita's love after kidnapping her.[134] Even this plurality of digital worlds and an infinite number of 'replays' cannot undermine Sita's essential nobility, however, and Ravana's attempt fails. Another story in the same collection uses Hindu speculation to revise what transcendent AI might look like in the future: an AI named Sita departs Earth and plans to self-replicate in its exploration of the cosmos.[135] This clearly imitates the end of the *Ramayana*, in which Sita bears children in her forced exile to the forest, though it empowers Sita (the AI) and recasts exile as voluntary. The author reasonably concludes that a powerful AI constructed in India might have different motivations from one created elsewhere: the AI takes inspiration from Hindu myth and makes active choices about its future on that basis.

Indian science fiction draws on local resources to resist apocalyptic dreams also. In the short story 'Icarus', for example, Subhojoy Gupta suggests that a hivemind supported by cyborg enhancements would bring about immortality and a scarcity-free economy but that it would stultify humanity.[136] Each human being becomes merely a computational resource—precisely what Moravec glorifies in his Mind Fire. Ultimately, an astronaut's unexplained cosmic vision gets transmitted to everyone through the hivemind, resulting in people rejecting their role as computers and encouraging a return to art, storytelling, and individual diversity.[137] It is hard not to see the influence of India's historical development here: many political and cultural actors point towards the potential for cultural pluralism and the richness of craft and culture in India's history.[138]

Alongside science fiction, the explosive popularity of smartphones among the middle and upper classes (and even among the economically disadvantaged) provides a clear conduit for transhumanist ideas in India. A man and woman whom I met in 2019 became friends thanks to her

---

[134] Dawesar, 'The Good King', 50–1.

[135] Das, 'Sita's Descent', 105–14, 113.

[136] Gupta, 'Icarus', Kindle location 303, 308.

[137] Gupta, 'Icarus', Kindle location 359.

[138] For two clear examples of Indian pluralism, see Nehru, *The Discovery of India* and Khilnani, *The Idea of India*. On India's craft legacy, see especially Chattopadyay, *India's Craft Tradition*.

posting about transhumanism on the website Quora. 'It's very difficult to find people' she told me, but the intellectual conversation enabled by their mobile phones led to a strong friendship by the time of our meeting. Such phones also increase access to global entertainment, which increasingly moves towards themes drawn from Apocalyptic AI.

In 2012, undergraduate and graduate students at IISc cited the Wachowski siblings' seminal film *The Matrix* (1999) as an inspiration for thinking about AI and technology, though neither it nor other science fiction films established a strong current of transhumanism in Indian science.[139] If such films cracked open the door for a more widespread interest in science fiction, smartphones have pushed it wide. Students and other young people now use Netflix and Amazon Prime to stream television and movies, and transhumanist students on IISc's campus explained that shows like *Black Mirror* and *Altered Carbon* provide direct access to western ideas of posthumanity and AI. In 2019, one reported that he found transhumanist inspiration:

> Mostly in the Hollywood movies. And then there is a limitation to that … I can imagine all this consciousness being uploaded into the computers and this neural lace whatever Elon Musk thinks he is doing, this superintelligence … this comes from books and movies. But I still think there used to be times when people used to think of things: what else can people think of these things. But it mostly comes from these Instagram and these things and I think it's still broad; that wall has to collapse. I think it needs to collapse and we need to take it much further… When Netflix comes, and then there are shows like *Black Mirror* … and you can afford it … In India there is also very solid concept of soul also. So still people are figuring it out … I feel like mind-body is still one.

Despite fear that outside IISc campus such ideas would be hard to discuss, he felt compelled to make connections where possible, and indeed it was a mutual friend who introduced us. Echoing his difficulty, another South Indian transhumanist told me that 'the buzz around A.I. and Netflix shows gives some exposure to the urban population on futurism

---

[139] See Geraci, *Temples of Modernity*, 147–56.

and transhumanism. However, I do not think that many of them form a deeply thought opinion on this issue or organize to discuss these ideas in many forums'.[140] We must therefore conclude that Indian transhumanism is in formation. Thanks to the network access provided by smartphones, transhumanist views of AI and human destiny have gained footholds in academia, industry, and entertainment.

India's cultural shift is not limited to pop culture; it includes small communities that foment quasi-religious thinking about technology and the arrival of intelligent AI. One of Bangalore's leading social innovators, Archana Prasad,[141] founded the art collective Jaaga in part based on a 'model of a temple to the Internet god, and the coming of sentient Internet'.[142] Describing her vision of godlike AI she told me:

> If I were an Internet god looking to reveal myself ... I would do that in India, and where in India except Bangalore ... the population, the conditions, the craziness of everything. How extremely urgent things are here, like in terms of crises ... there's no logic to this [laughing] ... India is a fertile ground for spirituality of any kind, and for religions new and old to take root. And so if there was going to be a temple to the Internet, where else would it be, honestly, and if it had to choose a very specific location, why not in the center of Bangalore? The initial devotees—all of our techies—are here, the possible converts.[143]

In saying this, she laughed with pleasure, indicating a more lighthearted engagement with such ideas than one gets from Moravec or Kurzweil. Unfortunately, Prasad's effort to detail some of her philosophical and religious ideas got truncated when she switched blog platforms, but one can still read some of her thoughts on posthuman evolution.[144] She writes that

---

[140] J. Prasad, email correspondence with the author, 6 December 2019.

[141] A. Prasad co-founded the international award-winning hacker-art collective Jaaga and has been selected as 'one of the 25 young people of importance to Bangalore city' (Team YS, 'Archana Prasad, Founder & Director, Jaaga'). For an example of Jaaga's significance, see Kamar, 'Community Space Wins Global Acclaim'.

[142] Interview with the author; see Geraci, *Temples of Modernity*, 151.

[143] See also Geraci, *Temples of Modernity*, 151–2.

[144] A. Prasad confirmed the lost blog content in an email to the author, 27 July 2015.

From the advent of homo-sapiens emerged a series of technological creations that will continue onwards eventually leading to a species that is faster, stronger, more capable than the homo-sapien in thought transmission and documentation. This is heralded by the Internet Era that we are now in, and is the next big milestone in our philosophy. We propound that all advancements are towards The Enlightened Singularity. This self-aware and all-aware nucleus, the creative whole is the guiding force for all occurrences, instances, manifestations, of the desire and will to create. Like traditional theism's [sic] this aspect of our philosophy remains unsubstantiated and is in fact a 'belief' system. This nucleus (as yet incomprehensible) is our 'God'.[145]

Archana Prasad's work with Jaaga revolves around more pedestrian concerns than emergent gods, but the inspiration and community building could provide a nucleus for growing interest in the impact of AI on our future.

Other small groups have formed in India, though not all have sustained their activities. The Bangalore Futurists, the India Future Society (IFS), and India Awakens are three such groups, with the latter two notable for their connection to western transhumanism. The IFS has Natasha Vita-More, James Hughes, and Aubrey de Grey on its advisory board. India Awakens is linked to The Turing Church (a group founded by Giulio Prisco, a leading spokesperson for transhumanism in the West who increasingly finds connections with Indian thought). While no such group has gained a large public following, none has yet the long years of engagement that would mirror transhumanism's slow but steady growth in the U.S.

Perhaps the most unique formulation of Apocalyptic AI is the Singularity Café, founded by Jayan Prasad in 2019. The café, which is in Chennai, serves coffee with a side of Singularity theory. Prasad produced three videos on the café and what he calls 'Impact Theory', all of which are available on YouTube. Prasad believes that if AI were programmed with a neural network that optimized its impact on the environment, it would in time develop into an artificial general intelligence (AGI) and

---

[145] A. Prasad, 'Enlightened Singularity'.

superintelligence. In the videos, he echoes the optimism of Apocalyptic AI, asserting that:

> 'They'll be able to be empathetic towards us ... It's going to do all the work. You know, solve the world's food crisis'.[146]
>
> 'All our needs will be taken care of by AGI'.[147]
>
> AI will treat us as pets or retired parents.[148]
>
> 'We will be happy anyways ... We [will] have accomplished our destiny by creating superintelligence'.[149]
>
> 'Your life will be more entertaining and more amazing. You'll probably have a flying broomstick and a magic wand ... magical devices AI will probably create and make your life better'.[150]

He adds to these fairly typical enthusiasms a novel approach: commerce and community.

> 'We should all collectively feel ownership of the thing we are creating ... You should just buy something ... if you contribute to this company that's actually creating AGI and that's creating the future generation, you should feel happy about the contribution you have done to bring out this impact ... you should see this as our child ... we have to look at AI as our child which has made it big in Hollywood ... Just as you love your [family members and community] AI will feel some sort of attachment to us'.[151]
>
> 'So what I would suggest ... do whatever you keep doing, you wake up and you go to work ... Come and contribute to a cause that is going and accomplishing something bigger ... creating AGI. That's human, mankind's destiny, I would say. You are contributing a little bit to achieve mankind's destiny. That's it. So you will be happier every day. So I would ask you to just come hang around in the café every day'.[152]
>
> 'Socialize with likeminded people, find your tribe here'.[153]

[146] J. Prasad, 'Singularity Cafe Chennai Part 3', 12:20.
[147] J. Prasad, 'Singularity Cafe Chennai Part 3', 13:50.
[148] J. Prasad, 'Singularity Cafe Chennai Part 3', 12:40.
[149] J. Prasad, 'Singularity Cafe Chennai Part 3', 14:20.
[150] J. Prasad, 'Singularity Cafe Chennai Part 3', 14:50.
[151] J. Prasad, 'Singularity Cafe Chennai Part 3', 11.
[152] J. Prasad, 'Singularity Cafe Chennai Part 3', 18:30.
[153] J. Prasad, 'Singularity Cafe Chennai Part 3', 19:15.

'Patronize the company that is contributing to AGI'.[154]

As proceeds from Singularity Café will support Prasad's own research in neural networks (he holds an advanced degree in this area and wants to build a research lab), the café doubles as a social hub and, at least in principle, a tech incubator. He has told me that even among his tech friends and family members employed as engineers that people 'do not get the vision for the future ... as far as they can imagine, things are going to continue to be similar to what it is now'. But he reiterated his hope that the café 'will eventually become such a place where people came and discuss these topics'.[155]

These groups are not neutral conduits for Apocalyptic AI; rather, they translate and transform Apocalyptic AI into their context. Through their connections outside India, they then provide opportunities for the reformation of western ideas (as seen in Prisco's work[156]).

For example, the founder of India Awakens, Nupur Munshi, argues that the transhumanist rejection of death and decay aligns it with the worship of Durga, the goddess who triumphs over evil. After describing the worship of Durga as an inspiration to conquer evil, she writes: 'I, a woman myself, bow to the Mother, that resides in all beings in the form of wisdom, power and compassion, to bless me with courage and strength as I move on with my colleagues in Turing Church in our endeavor to bring back all those who have left us ... for an eternal life together'.[157] Munshi, who also sees human-equivalent AI as part of the tradition of 'building gods' in Indian religious practice,[158] draws on the traditions and worship of Durga to justify her transhumanist vision of salvation.

Even Jayan Prasad, who speaks openly of his atheism, will probably find that if Singularity Café succeeds the versions of transhumanism that flourish there will attend to India's cultural and religious values. As more Indians become aware of transhumanist promises and seek to find common ground between those promises and their traditional

---

[154] J. Prasad, 'Singularity Cafe Chennai Part 3', 19:25.
[155] J. Prasad, email correspondence with the author, 6 December 2019.
[156] Prisco, Tales of the Turing Church, 79–89.
[157] Munshi, 'Turing Church Manifests Durga's Symbolism of Good over Evil'.
[158] Munshi, 'Building Gods'.

cultures, we will see the growth of transhumanism in India, but also its translation—new idioms will produce new directions and areas of focus.

The cultural shifts visible in entertainment and in religious specula- tion also appear in Indian science. At a policy level, India's technology roadmap for information and communications technologies suggests that machine intelligence will soon match human intelligence. The report claims that 'this watershed in the history of computers and of the human race will occur at some point between 2030 (according to the optimists) and 2045 (according to the pessimists).[159] For obscure reasons, this docu- ment provides 2045 as a pessimistic date even though interviews with AI experts provide a much wider range for such potential outcomes.[160] Intriguingly, it appears that while AI experts in India do not hold the most optimistic timeline for the next 20–25 years, they are the most opti- mistic of all nations' researchers in their faith that human-level machine intelligence will be achieved after that period.[161] For the government's technology innovation roadmap, machine superintelligence is not some- thing to be feared but instead a 'technological miracle' allowing humanity to 'benefit from superior artificial intelligence.[162]

While Wadhawan's early transhumanist cheerleading in *Resonance* went unheard, transhumanist speculation now takes a more substantive position in India's pop science literature. The magazine *Dream 2047*, which is pub- lished by Vigyan Prasar (an 'autonomous organization under Department of Science and Technology' of the Indian government[163]), only rarely engaged in futurism during its opening decades but as of 2018 took an increasing interest in such ideas. From its launch in August 1998 through December 2018, *Dream 2047* published only five essays and three short news items that engage in at least some futurist speculation.[164] Out of these publications, three of them appeared in 2018, meaning only five such pieces

[159] Technology Information, Forecasting and Assessment Council, *Technology Vision 2035: Technology Roadmap on Information & Communications Technology*, 73.

[160] See Grace, et al., 'Viewpoint: When Will AI Exceed Human Performance?'

[161] Ibid., 738.

[162] Technology Information, Forecasting and Assessment Council, *Technology Vision 2035: Technology Roadmap on Information & Communications Technology*, 73.

[163] Vigyan Prasar webpage.

[164] Misra, 'Shaping the World Atom by Atom;' Singh and Chaturvedi, 'Dreamer and Thinker;' Tripathi, 'Recent Developments in Science & Technology;' Basu, 'Recent Developments in Science and Technology' (two short pieces); Maitra, 'Demystifying the Human Brain;' Das, Dey, and Banerjee, 'Crisp and CRISPR;' Bhattacharjee, 'Age of Man-Machine Hybrids'.

were published in the journal's first 19 years. Meanwhile, a host of articles engage relevant technologies, such as computers, nanotech, and genetic engineering without applying a transhumanist perspective to them.[165]

Overall, *Dream 2047* shows profound interest in space science, scientific biographies, and human health. It includes year-long commitments to particular technologies or sciences. But over its first two decades, it was surprisingly inattentive to robotics, AI, and information technologies in general (despite the massive growth in India's IT industry during that time). In the years surveyed, the journal contains nothing on how computers function, how mobile phones or their networks function, nothing on robotic cars, and nothing on recombinant DNA until CRISPR Cas9 was invented.[166] Certainly, pop science authors are under no obligation to align with or even express transhumanist interests, but the absence of those viewpoints and the technologies that ground them seems surprising in an era where Kurzweil became a bestseller and everyone following him in the U.S. felt obliged to at least acknowledge the Singularity, human enhancement, or intelligent AIs.

The more recent *Dream 2047* articles that explore transhumanist speculation pay closer attention to western futurism. In 'Age of Man-Machine Hybrids', Govind Bhattacharjee avers that based on 'the exponential growth of technology, it is only a matter of time before [robots] will acquire self-awareness that rivals human intelligence'.[167] Bhattacharjee's agreement with western futurists does not stop there:

> There is still a long way to replicate human intelligence, but it may usher in an era in which there will probably be no such thing as 'pure' human intelligence, because all humans will be a combination of biological and non-biological systems which will constitute integral parts of our physical bodies, vastly expanding and extending their capabilities.

---

[165] Sehgal, 'Seminar on Information Technology in India in the Next Millennium;' Kamble, 'Making Science More Accessible and Less Frightening;' Kamble, 'Clones—It's Human Beings Now;' Salwi, 'National Centre for Software Technology;' Behera, 'Science and Nanotechnology of Nanomaterials;' Khened, 'Gordon Moore, His Law, and Integrated Circuits;' Mishra, 'Nanoscience and Its Applications;' Mohanty, 'Nanotechnology in Environmental Remediation;' Murthy, 'New Tools for Gene Therapy;' Bhatia, 'CRISPR—Cas9'.

[166] I should note that the online archive of *Dream 2047* does have some broken links, especially in the year 2009. As a consequence, this evaluation does not include a dozen or so missing issues.

[167] Bhattacharjee, 'Age of Man-Machine Hybrids', 30.

Humans and machines will merge together to create a human–machine civilisation.[168]

Unsurprisingly, Bhattacharjee relies heavily on western futurists in coming to this conclusion. He cites extensively from John Brockman's edited collection *What to Think about Machines that Think* and Kurzweil's *The Singularity Is Near*. Bhattacharjee considers machine consciousness and what it will mean for us to live in a future with intelligent machines and augmented human beings. He states that 'humans and machines will merge to constitute a unified entity where the distinction between man and machines will be obliterated. The question itself whether machines can have consciousness will then become meaningless'.[169] In another essay, Bhattacharjee rejects terminator scenarios, writing that 'future machines may look as if they really have a "soul," but not one that will destroy. Instead it will build, and aid us in making the human condition a little better'.[170]

Following Bhattacharjee's powerful support for the basic claims of Apocalyptic AI, a 2019 article in *Dream 2047* continues the magazine's interest in AI and transcendence. Using AI victories over human gaming champions as evidence, Deepak Kohli argues that machines will soon be more broadly intelligent than humanity: 'fortified by AI, Google's AlphaGo can defeat human player [sic] in a computer board game called Go. This clearly shows that AI has the potential surpass humans'.[171] Kohli is not as sanguine as Bhattacharjee, however, and notes that 'AI may make machines free from the control of humans in future [sic], resulting in a clash between machines and men'.[172]

Whether through conflict or mutual growth, futurists see transformative potential in AI, and it is here that the science of AI slides so easily into religious or quasi-religious predestination. Such views are not limited to the U.S. but can emerge in Indian contexts also. Already I have noted Archana Prasad's vision of 'Enlightened Singularity' and Nupur Munshi's

[168] Bhattacharjee, 'Age of Man-Machine Hybrids', 29.
[169] Bhattacharjee, 'Age of Man-Machine Hybrids', 28. Note that the essay is reverse-paginated. This claim is Bhattacharjee's conclusion.
[170] Bhattacharjee, 'The Curious Case of Unruly Robots', 19.
[171] Kohli, 'Artificial Intelligence', 28.
[172] Kohli, 'Artificial Intelligence', 28.

alignment of Durga with advanced technology. As quoted above, a student attending a lecture I delivered in 2019 to students and faculty at Srishti Institute of Art, Design and Technology raised his hand at the end to ask whether AI could be Kalki, the final avatar of Vishnu that is predicted to end *Kali Yuga* and inaugurate a return to *Satya Yuga*.

Kalki either has the head of a horse or else rides a horse. Sacrificial horses being vital to cosmic maintenance and reconfigurations, it should come as no surprise that the eschatological transition of cosmic history would involve a horse in Hinduism. In his science fiction books, *Turbulence* and *Resistance*, Samit Basu riffs off the religious anticipation of Kalki's return: as part of a mutation event that produces superhumans, a baby is born with a horse's head and that baby grows up as Kalki. In *Resistance*, various actors seek to achieve a posthuman future at the expense of the majority whom they would relegate to nothing more than breeding stock.[173] Kalki, however, resists these machinations—including an effort to induce a posthuman future of giving all people superpowers—and creates a new world that humanity can populate and fashion in fruitful ways.[174] The new world sought by would-be posthuman rulers fails. Kalki, not coincidentally integrated with a computer system that leverages his world-altering powers, achieves his own godhood and fulfills the promises of religion: he launches a new world where people can begin anew.

Basu's stories perfectly illustrate the plausibility of foretelling different futures in an Indian context. Indian transhumanism and Indian approaches to AI need not follow Kurzweil's prophetic imagination. Basu decides against a world where all humanity transcends to posthumanity, but he nevertheless acknowledges the plausibility of new worlds and new opportunities. As such, *Resistance* ends with a transhumanist vision of the future even though Basu rejects the mainstream libertarian transhumanism that runs rampant in the West.[175] Traditional mythology may

---

[173] Basu, *Resistance*, 218–20. In this usage, 'posthuman' refers to a technologically evolved species, not the literary or philosophical criticism of humanism.

[174] Basu, *Resistance*, 285–9. The book's characters believe that permanently giving superpowers to all humanity was simply not possible for Kalki (p. 283), but given Kalki's apparent apotheosis (p. 284), it is unclear whether we are to read this as Kalki's failure, the impossibility of posthumanity, or Kalki's decision to fulfill the letter of the request but not the spirit.

[175] Basu is not the only source to which we might turn for an ideological difference between India and the West. In his article on Tagore—and certainly not with reference to transcendent AIs but still with an interest in evolving humanity—K. C. Sen suggests that the future 'God-Man' must be 'boundless in power and boundless in love' ('The Religion of Man', 258). He criticizes the oft-times hero of transhumanism, Nietzsche, who 'had power only and no love' (Sen, 'The

have prevented the explosive growth that science fiction experienced in the U.S., but contemporary Indians integrate the different cultures, blending religion, technology, and speculative futurism into their own literary, scientific, and religious worlds.

Conclusion

D.K. Wadhawan hasn't found an attentive audience, and it remains to be seen whether Bhattacharjee, Munshi, or another Indian thinker will fare better than he. Nevertheless, Apocalyptic AI shows clear growth in Indian society and its translation into Indian contexts opens the door for specifically Indian inflections of transhumanism and in our global visions of artificial intelligence.[176] The success of global futurist speculation will depend upon how effectively western perspectives connect to nonwestern worldviews,[177] and this would require transformative sacrifices. Either Indians must give up something of the culture that they have nurtured or western transhumanists must give up something of their own. Taking Bangalore as an example, one sees that policymakers, landowners, and industry willingly and disastrously sacrifice the city's trees, forests, and lakes in exchange for developing India's IT industry.[178] But the future need not come at the expense of Indian's cultural or natural wonders—it is well past time that western society make sacrifices towards a greater human community.

In moving beyond western dominated transhumanism, perhaps the neo-libertarian ideals common in U.S. transhumanism might be

---

Religion of Man', 258). Ultimately, suggests Sen, the God-Man must relinquish power in favour of love, a choice of sacrifice that appears absent from transhumanist thinking, where one need never make sacrifices in pursuit of individual fulfillment.

[176] This book refers to Hindu traditions, but those are not the only possible sources for Indian contributions to transhumanist aspirations. For example, there are fledgling efforts to engage transhumanism from Muslim perspectives (see Jackson, *Muslim and Supermuslim*; Hejazi, 'Humankind. The Best of Molds'), and the community of Indian Muslims might contribute to those.

[177] As noted in the introduction to this book, I am alert to the difficulties posed by terms such as 'western' and 'nonwestern'. While these are difficult to sustain against close scrutiny due to the very networks of interaction I discuss in this chapter, the terms nevertheless provide some shorthand for the expression of geopolitical differences.

[178] Nagendra, *Nature in the City, passim*, but especially pages 146, 173.

discarded.[179] Nonwestern values—such as the Korean value *xiang*, in which a human being is situated within a social/environmental context—could be a valuable tool in overcoming self-centred worldviews that prevent us from building technology to end oppression.[180] Cheering for free markets and individual decision-making serves as a magic trick: handwaving that distracts us from how real people get denied the possibility of economic and political freedom (specifically those who are non-white, nonwestern, and not men). This speaks to a larger concern that most thinking about technology—and artificial intelligence is a prime example—gets wrapped up into the concerns of entrenched power structures. Just as neo-liberal market economics too often leaves the marginalized no mechanism for joining the mainstream, transhumanist imagination often deliberately obfuscates the impossibility of impoverished communities gaining access to technologies of enhancement.[181]

Before further noting how western cultural powers must sacrifice the very structures that permit their dominance, it is worth noting how India can also transcend misguided approaches to technology, economics, and politics. In particular, faith in Vedic technology—the belief that ancient Indians long ago possessed all of modern science and technology described in Chapter 1—must be sacrificed to bring about a new global role for India. Such faith was perhaps helpful to the independence movement and was a mechanism for adopting science and technology into India: it promoted the indigenous, and hence worthy, nature of science and technology. But it presently opposes progressive scientific research; contemporary Indian scientists rightly declare faith in Vedic technology to be an

---

[179] Graham, 'Nietzsche Gets a Modem', 70; Tegmark, *Life 3.0*, 163–8; Clayton, 'The Ridicule of Time', 323, 329, 336. See also Pein's no-holds-barred attack on cryonics, which includes a criticism of its libertarian associations; 'Everybody Freeze!', 90–1 (I feel obliged to note that Pein's essay fails to cite its claims, many of which are attributed to well-known transhumanists, Silicon Valley entrepreneurs, and others; this failure to guarantee the veracity of one's source material is not inspiring).

[180] On Qiang and transhumanism, see Jung, 'Transhumanism and the Theology of *Xiang*'.

[181] For a variety of examples revealing transhumanists blaming the poor for their plight, suggesting or asserting that the poor cannot participate in transhumanist cultural and technological advances, and/or ignoring the global poor (despite occasional suggestions that the future will solve poverty), see Ettinger, *The Prospect of Immortality*, 125, 127; *Man into Superman*, 10, 137–9, 161, 249; Esfandiary [1973] 177, 114; FM-2030, *Are You a Transhuman?*, 32, 73, 75, 77. Natasha Vita-More makes a better effort to connect the needs of the poor to transhumanism; but by blaming all problems on authoritarian governments, she ignores the responsibility of neo-liberal market economics (see *Transhumanism*, 14).

attack on the very soul of India.[182] Having served a purpose in colonial India, it is time to let this belief die. Many scholars already note the real, historical contributions of Indian culture to science and technology[183]; Indians should turn to these rather than conflate mythology with history. The departure from Vedic technology could open new opportunities for thinking about science and technology. In the next chapter, some of the ramifications of this will be explored.

New forms of speculative literature could entirely forsake India's past, ditching the *Ramayana*, the *Mahabharata*, the *puranas*, and other myths in favour of disenchanted science fiction and pop science—but this is unlikely to be productive. Those who have forsaken tradition in favour of unfiltered transmission found little or no traction among Indians. One need not look very far, on the other hand, to find interesting science fiction or even living transhumanist communities that retain the influence of Hindu traditions while interrogating modern technology. Those who similarly translate and transform, taking advantage of their cultural heritage, are more likely to have success in the years to come.

Perhaps Indian transhumanism will drive away the last vestiges of a colonial mindset that prevents bidirectional transmission of ideas. In their evaluation of Samit Basu's fantasy-science fiction novel, *Simoqin Prophecies*, Nasima Mazarbhuiyan and her collaborators note the politics in literary analysis that lead towards western culture being the default position where western influences should be easily understood and in which Indian elements must somehow be explained. They rightly criticize this asymmetry and argue that the Indian elements should be on the same cultural and critical playing field as those of the U.S. and elsewhere.[184] This specific example speaks to the larger question of how to bring Indian (and other cultures') ideas to western nations through literature, science, politics, and more. Within the field of AI ethics, voices have emerged arguing that AI must be decolonized, broken from its moorings in unequal power structures that resulted from colonialism.[185] This perspective must

---

[182] Vahia, 'Evaluating the Claims of Ancient Indian Achievements in Science', 2148; see also Narlikar, 'The Scientific Edge', 1. I further engage the scientific opposition to Vedic technology in *Temples of Modernity* (pp. 82–5) and the overall political process of Vedic technology in 'Saffron Glasses'.

[183] For example, Ray, *A History of Hindu Chemistry*; Baber, *The Science of Empire*; Subbarayappa, *Science in India*.

[184] Bose, et al., 'Indianness?', 341.

[185] See Mohamed, Png, and Isaac, 'Decolonial AI'.

be applied across the cultural perspectives that govern technology, including digital technology.

The dramatic intersection between colonial white power and technological visions must be sacrificed to inaugurate a new world. Syed Mustafa Ali argues that western transhumanism is very white, very male, and very well positioned to maintain traditional hegemonies and global power structures that trace to Christendom's medieval battle with 'the infidel'.[186] Steven Cave and Kanta Dihal echo this, describing ways in which the portrayal of AI as white excludes non-whites from futurist imaginary and affirms social prejudices against them.[187]

Recognizing similar problems in transhumanist thought, Philip Butler offers a powerful critique of racism in humanist philosophy and its transhumanist manifestations while arguing that transhumanism could be made liberatory for the Black community.[188] The white supremacist ideology that helped shape the boundaries of the transhumanist kingdom must be asphyxiated like the horse in the *Ashvamedha* to inaugurate the newly formed world.[189] The apocalyptic future might be one enjoyed by all people instead of solely by white men. In keeping with this, it is intriguing that the founder of India Awakens, Nupur Munshi, ties her transhumanism to conquering evil. Those who describe and design AI must all strive actively against the evil of white supremacy. That struggle to bring about justice through technological progress will produce tangible benefits for the world.

The new religious movements of India re-imagine western transhumanism with more careful attention to their traditions; so despite the fact that many western transhumanists declare themselves ardently opposed to religion, they may find themselves accommodating a more

---

[186] Ali, '"White Crisis" and/as "Existential Risk."' See also Butler, *Black Transhuman Liberation Theology*, 139–40.

[187] Cave and Dihal, 'The Whiteness of AI'.

[188] Butler, *Black Transhuman Liberation Theology*. For the critique of whiteness in humanist and transhumanist thinking, see pp. 27–39.

[189] We might legitimately ask whether an ethical world can be founded out of an unjust one. Butler certainly seems to believe so, as do I. After all, what choice do we have? Butler moves past the awareness of injustice and its thorough penetration into the structures of contemporary life with a spiritual hope for the future (*Black Transhuman Liberation Theology*, 137–9, 140–4). In the *Ashvamedha*, the horse is a symbol of violence because its passage marks the threat of military power. Its sacrifice in the ritual is—symbolically, at least—a recognition that the new kingdom should be one that escapes the violent exercise of control. The horse, for all that its symbolic power traces the lines of the kingdom, simply cannot live on to enjoy that world.

openly religious vision of the future. The fact that Indian transhumanists do not necessarily give up their traditional gods or traditional practices in building their metaphors could have significant bearing on the future of transhumanism. What might be sacrificed in the development of global transhumanism depends on who has a say in producing visions of cosmic transformation. Scientists, engineers, artists, policymakers, philosophers, business leaders, consumers, and others will share this responsibility.

Simultaneously, we must ask whether that growth will happen collaboratively or whether each nation will forge its own path. Already the connection between India Awakens and The Turing Church has led Giulio Prisco to an increasingly and explicitly religious orientation. Where once Prisco described religious transhumanism as a helpful 'front end' to attract traditionally religious people to transhumanism, his own work now seems to have moved beyond matters of rhetoric and is thoroughly infused with this spirit.[190]

Assuming that we can build intelligent machines, their motives will surely vary by cultural context. The idea of one machine culture is as implausible as the idea that all humanity will unite under one set of laws, rules of etiquette, and social perspectives. In any case, transhumanism and the apocalyptic views that have become widespread in conversations about AI are now subject to their own evolution. The Euro-American visions that fostered Apocalyptic AI cannot suffice alone even for the western world: they are too limited and haunted by unjust power structures. Certainly, those visions are deeply inadequate in India, not to mention China, Ethiopia, or any of a host of nonwestern countries we might consider. Nonwestern transhumanists will continue taking hints from westerners and possibly these latter will even retain a majority stake in the intellectual leadership of global transhumanism. But Indian transhumanists will imagine new speculative futures, and these will enliven global transhumanist thought. Simultaneously, they can help

---

[190]   This conceptual arc is clearly visible in Prisco's published writings; see Prisco, 'Engineering Transcendence;' 'Transcendent Engineering;' *Tales of the Turing Church*. His relationship with nonwestern thinkers may be decisive in this transition, though his explorations in *Tales of the Turing Church* (pp. 30–1) indicate that he may have held to a strongly religious transhumanism even while describing it as a rhetorical move to transhumanists who resist the religious interpretation.

shape the design and deployment of AI technologies, which need not be restricted to Singularity aspirations and uploaded immortality.

In the next chapter, we will build on the importance of global AI discussions by considering how Indian values can contribute to a specific and pressing domain of AI research: surveillance. Artificial intelligence technologies greatly enhance military, political, and economic control over civilians. Managing the threats posed by surveillance is relevant to the immediate present but also to the future development of AI. The transhumanist dreams that have occupied these past two chapters cannot be separated from the practicalities of everyday technology. If, for example, human- or greater-than-human equivalent machines arise out of a surveillance regime dedicated to the manipulation and control of human beings, we can count out any possibility of the positive transformation promised by Apocalyptic AI. Therefore, we must consider how values outside of Apocalyptic AI can be leveraged in AI design, which can thus render the future open to possibility. Just as Hindu traditions can be used to rethink the Apocalyptic AI agenda, they can also open new models of reflection in conversation with it.

4

Recoding Religion

Theological Renderings of AI and the Future of Society

Introduction

The arrival of apocalyptic perspectives on artificial intelligence (AI) in India provides us with an opportunity to consider how our technological imaginaries can be used to produce a better future for humanity. The translations and reconfigurations noted in the prior chapter offer clear evidence that Apocalyptic AI will not simply dictate how Indian scientists, technologists, or average citizens think about AI. This means that India and other nations have an opportunity to engage in dialogue with western thinkers and contribute towards a global vision of AI. This chapter takes a specific domain of AI, its capacity to observe and control human behaviour, and argues that both American and Indian value systems can be profitably harnessed towards a better future.

Apocalyptic AI offers little firm protection from social ills. Kurzweil recognizes some problems we might face, but breezes past them, suggesting that we will stay out in front of the difficulties with other technological fixes.[1] This assumes, of course, that problems scale at the same rate as solutions—a dubious proposition given the array of problems we already face. As mentioned in the introduction, AI technologies contribute fundamentally to overthrowing individual sovereignty and personal responsibility, dubious banking and judicial outcomes, cars that cannot see darker-skinned people, and other social problems. The previous chapter noted how whiteness gets encoded in AI and robotics and argues that we must rebuild the world by sacrificing unjust social structures. We can legitimately ask how dreams of the Singularity and mind uploading will

---

[1] Kurzweil, *The Singularity is Near*, 391–426.

Futures of Artificial Intelligence. Robert M Geraci, Oxford University Press. © Oxford University Press 2022.
DOI: 10.1093/oso/9788194831679.003.0005

solve these concrete problems. Meanwhile, political actors have reason to ignore the problems or even to directly exacerbate them. The problem of surveillance, which occupies this chapter, illustrates both the immediate need for collective engagement with AI and the insufficiency of Apocalyptic AI as the guiding principle for advancing our collective future. Rather, the problems posed by surveillance demand a global effort at problem-solving.

It is easy to imagine AI as a panopticon: a surveillance system that leverages fear of observation to habituate regimes of control (especially self-control). The panopticon was designed by Jeremy Bentham as an architectural solution to information control in buildings with large populations requiring observation and control: prisons, hospitals, and schools. In its primary incarnation—the prison—the panopticon design placed a warden in a central tower with slot windows. From that vantage, the warden could conceivably observe every inmate without them knowing whether or not he was looking at them at any given moment. The architectural possibility of observation—with the looming threat of punishment—created a psychological demand on the prisoners to act as though the warden was observing them at all times. Bentham's idea was that prisoners could thereby be made to self-regulate.[2]

Facial recognition and pattern recognition technologies employed at airports, sporting events, and even city streets work precisely towards the panopticon's goal. Implementation of such technologies is underway despite insufficient public debate and even less oversight.[3] Other AI systems aim towards control by motivating consumer or social behaviours. There is strong evidence that such systems have an effect on voting behaviour that is deleterious to democracy.[4] Eventually, AI systems might supervene on policy matters or, if they reach the transcendent heights promised by their promoters, even subjugate humanity. Differentiating among such outcomes depends upon more than simply technical capability. Cultural ideas—in the present case those of religion—are part

---

[2] For a broader description of Bentham's design and its influence on regimes of control, see Foucault, *Discipline and Punish*, 200–10.

[3] For example, see Alba, 'The US Government Will Be Scanning Your Face at 20 Top Airports, Documents Show'.

[4] Grassegger and Krogerus, 'The Data that Turned the World Upside Down;' Gallacher, et al., 'Junk News and Bots during the 2017 UK General Election'.

of technological conceptualization and deployment. As such, attention to different cultures' religious perspectives could open doors for intentional engagement in the future of how AI systems produce or reinforce social order.

The purpose of a panopticon is 'to induce in the inmate a state of conscious and permanent visibility that assures the automatic functioning of power;'[5] it might thus be compared to other control systems, such as the superego (as described by Freud) or the governor in Watt's steam engine. For all three, a condition that might be expected to run amok is constrained by a system that checks for out-of-balance conditions and rectifies them without external intervention as long as the system is functioning. The governor controls the flow of steam. The superego restrains socially unproductive impulses.[6] The panopticon deters counter-indicated behaviours from prisoners, students, etc. The power of observation is central to all these systems of control and the two human examples both rely on the presumption that an individual can be under observation at any time without knowing it. For both Freud and Bentham, it is the expectation that one could, in principle, be observed at any time that forces self-control.[7] In the steam engine, a machine does, in fact, constantly observe the situation through its own mechanical affordances. AI promises a unification of human self-regulation and machine observation. Surveillance never ends, thus intensifying the human experience of being observed and subsequently the human effort to self-regulate.

The global development of modern science has created complex and interconnected networks through which both technologies and ideas circulate. Within these networks, AI technologies expand their influence across human life, thus establishing structures that define human thought and practice. If AI should develop human-equivalence or greater-than-human intelligence, then the influence of AI on society will be even more profound. Such technologies do not develop in a cultural vacuum, however, but are the result of pre-existing cultural and religious practices. In

---

5 Foucault, *Discipline and Punish*, 201.
6 Freud, *The Ego and the Id*, 18–29.
7 It is worth noting that Freud's highly dubious theory of the origin of religion relies upon the individual's presumption of being observed by spiritual entities (*Totem and Taboo*, 177–82). In *God & Golem, Inc.*, cyberneticist Norbert Weiner first sought points of contact between religion and AI worth considering through the question of control systems.

general, as Mackenzie and Wajcman note, 'the characteristics of a society play a major part in deciding which technologies are adapted'.[8] As a result, religious and cultural traditions will have definite impact upon the global development of AI even as those technologies then define how people will live and work in the future. Science fiction and popular science approaches to AI in the U.S. and India demonstrate that human intervention in AI development is possible and that there are no predetermined outcomes in how AI might further develop and subsequently structure society. Apocalyptic religious ideas underlie perspectives on AI in the U.S. and lend themselves to transformative dreams of aspiration and hope but also enable values of capitalist and military domination. By contrast, recent Indian perspectives on AI leverage religious structures of duty that could become tools for collaboration and care, or could reify a new caste system. Which religious and cultural perspectives we employ as we build increasingly powerful machines and which stories we tell about such machines will dictate, in turn, how those machines constrain and control our own decision-making and our future lives.

The undetermined future

Popular and even technical discussions of technology often presume the inevitability of specific future outcomes. Indeed, conversations about computers and AI are almost exclusively described in those terms. The apocalyptic visions of technology outlined in Chapter 2 are a dramatic example of this, though they are not without their critics. The allure of technological determinism, however, goes beyond AI and occupies a wider and stronger place in late 20th- and early 21st-century perspectives. It is therefore necessary to expose the fissures in our ideological commitment to technological determinism, opening both the present and the future to critical consideration.

It is common for people to ignore the past and see technological developments through a logic of determinism; in part, this is because we ignore what never managed to dominate the marketplace or the technological landscape. We pay more attention to successful technological

---

[8] Mackenzie and Wajcman, 'Introductory Essay', 6, emphasis removed from original.

developments than unsuccessful ones, leading us to the presumption that 'the success of an artefact is an explanation of its subsequent development'.[9] We then believe that technologies uniquely produce results in the social world and thus the next stages in their own development: that technological progress is defined entirely by itself, by its own inner workings, and that it offers its own guarantees for the future.

Sometimes, one choice necessarily excludes other choices and constrains future development. This phenomenon, path dependence, often dominates engineering decisions through internal feedback mechanisms (such as proofing successor technologies against incompatibility with precursor variations) and external feedback (such as dictated by aesthetic choices or political involvement). It is, however, tautological and thus uninteresting to assert that technologies as they are provide technological constraints on technologies to come.

Path dependence should not be assumed to necessitate technological determinism, only that the present is relevant to the future. Further, the fact of path dependence should not be presumed to prevail across shifting technological paradigms. The existence of QWERTY keyboards on a typewriter may have relevance to computer keyboards, but it has less obvious impact upon hypothetical brain–machine interfaces. It may be that the reign of QWERTY will continue to daunt its opponents in a transhuman future of cyborgs and virtual worlds, but there is no necessary reason to believe it will.[10] Path dependence is a very real issue, however, in the development of AI;[11] interrogating such processes can open opportunities for progressive engagement.

We tend to see technological change as a process of linear and obvious developments—after the fact, and with all the controversies, questions, and concerns absorbed into the accepted technology—but human factors are a necessary part of technological development. Combined with the obvious impact of path dependency, the linear view of technological development encourages a belief in technological determination. But

⁹ Pinch and Bijker, 'The Social Construction of Facts and Artefacts', 406.
¹⁰ For an exemplary complaint against the supposedly illegitimate domination of QWERTY keyboards, see Jared Diamond, 'The Curse of QWERTY;' for a slightly less emotionally charged description, see Dorit, 'Marginalia: Keyboards, Codes and the Search for Optimality;' finally, for a review of the debate and a computational effort to resolve it, see Torres, 'QWERTY vs. Dvorak Efficiency'.
¹¹ Rahwan, et al., 'Machine Behavior', 481.

human beings maintain agency in this process, even if we are inclined to believe otherwise. Human choices produce technologies, and therefore the ends to which technologies aspire are implicated in human society.[12] In their analysis of technological artifacts, Pinch and Bijker offer an excellent example: the dumping of war supplies in 1918 led to more phenol (carbolic acid) in the marketplace, which produced a drop in cost, which made Bakelite affordable and economically competitive.[13] Social dynamics include government engagement, international politics and war, the rate of technological adoption across geographic space, across industry, and the home. All of these and more affect whether and how a technology develops, and how its use unfolds.

Consider, for example, the role of the *charkha* (spinning wheel) in India's independence movement. British colonialism undermined India's dominant role in global textiles[14] and produced widespread economic privation through its taxation and trade systems. Gandhi then advocated dismantling industrial textile production in favour of individual cotton spinning.[15] While Gandhi's desire for universal village independence from national and global marketplaces and modes of production was impractical, the *charkha* nevertheless became a tool in the nationalist repertoire. Two things are worth noting. First, the 20th-century usage of *charkha* may never have gained any significant currency and certainly wouldn't have become a political symbol, without British colonial oppression. Second, that national politics were partially determined by the existence of the tool. Ultimately, these complex dynamics resist reduction to simplistic narratives of technological determinism.

Technological determinism is a form of crypto-religion mixed up in Euro-American views of divine providence. In *The Religion of Technology*, David Noble shows how the emergence of modern technological production in Europe was directly tied to religious goals and was ideologically suffused with the idea of divine providence.[16] Many viewed technological progress as the fulfillment of the Christian god's plan for the future. The

[12] Feenberg, 'Subversive Rationalization: Technology, Power, and Democracy'.
[13] Reviewed in Pinch and Bijker, 'The Social Construction of Facts and Artefacts', 406.
[14] See Adas, *Machines as the Measure of Men*, 41–53; Parthasarathi, *The Transition to a Colonial Economy*; Nair, *The Promise of the Metropolis*, 41–3.
[15] Gandhi, *Indian Home Rule*, 107.
[16] David Noble, *The Religion of Technology*.

correspondence between technological outcomes and divinely ordained history suffused technology with an ideology of historical providence that persisted even after technology had been divorced from direct association with Christian millenarian history. Claims that AI must inevitably evolve into a specific future form are therefore contingent upon a false logic. But that does not mean that those technologies and their futures will not play out like a Greek tragedy, with the prediction producing the performance.

Previous chapters discuss the aura of inevitability that suffuses Apocalyptic AI, but of course that aura is not unassailable. Kurzweil believes it possible to predict future technologies based upon current trajectories, but his record here is much less successful than often alleged.[17] Regardless of specific predictions, however, the overall thrust of Apocalyptic AI is to declare that digital technologies unfold according to either rules specific to computers, such as Moore's Law and any future iteration of it, or rules general to life or the universe.[18]

In contrast, Katherine Hayles argues for the contingency at stake in the history of computing and demonstrates how active decisions shaped computing while excluding alternative visions.[19] If human choice mattered in the historical past of computational technologies, then it surely matters in the future—regardless of apocalyptic proponents' claims to the contrary. In his novel *Seveneves*, Stephenson defines 'Amistics' as the study of 'the choices that different cultures made as to which technologies they would, and would not, make part of their lives'.[20] In his telling, both human enhancement and digital technologies are subject to such choices.[21] Given

---

[17] Kurzweil, *The Age of Spiritual Machines*, passim; *The Singularity Is Near*, passim. Kurzweil, himself, claims a tremendous success rate for his predictions (see Hochman, 'Reinvent Yourself'), but his own interpretations of success tend to be generous. For a nice summary of Kurzweil's predictions in *The Singularity is Near*, see Feakins, 'The Singularity is Near'. Feakins tallies up Kurzweil's predictions and evaluates Kurzweil's success rate as of 2017, which he lists as 9 out of 25 (that could be definitively judged as of 2017). It is worth noting that among the nine correct predictions, several are decidedly pedestrian and were made by a wide swath of commentators, and others require a fairly generous interpretation to be labeled true. None of the more outrageous predictions (barring perhaps effective language translations) have come true. For even more critical perspectives, see Knapp, 'Ray Kurzweil's Predictions for 2009 Were Mostly Inaccurate;' Rennie, 'Ray Kurzweil's Slippery Futurism'.

[18] Moravec, *Mind Children*, 100; Kurzweil, *The Singularity is Near*, 7–14.

[19] Hayles, *How We Became Posthuman?*; Hayles, *My Mother Was a Computer*.

[20] Stephenson, *Seveneves*, 611. The term is drawn from the Amish, a Christian denomination known to eschew much modern technology.

[21] Stephenson, *Seveneves*, 611–2, 640–2.

the possibilities that once existed in thinking about computers, it seems trivially obvious that the future remains open to intervention. Alternate visions of technology are in part dependent upon religious and cultural contexts. While Kurzweil sees AI as crucial to personal immortality, one of Japan's most renowned roboticists, Masahiro Mori, relied on his Buddhist experience of non-duality to spur technological creativity.[22] These differences clearly imply that religious perspectives are important to technological Amistics.

Amistic choices and the cultural values behind them reflect on the questions of social control that drive this chapter and suggest that we can consciously use AI to promote fruitful outcomes. Within the realms of computer science and cybernetics (the study of control systems in human beings and machines), there are clear advocates for human freedom. Joseph Weizenbaum, a seminal figure in the history of both fields, writes in *Computer Power and Human Reason* that technology could not be self-determined and was instead driven by human interests and concerns.[23] As an example, he argues that scientists designed the McNamara line because they opposed the bombing of North Vietnam and thereby contributed to the rise of the electronic battlefield. Rather than technological determination, there were people choosing how to behave and what to do; they could have done otherwise, he argues, such as by opposing the war altogether.[24] He raises this example precisely to point towards the importance of good decision-making in the development of computing technologies. Similarly, 21st-century AI could be driven by values outside those of the military-industrial complex or the commercial interests of surveillance capitalism.

Ideologies and technologies

We build social structures and ideas into technologies, so the apocalyptic claims that captivate public imagination have clear social relevance. In a simple and perhaps obvious example, the individuals who prognosticate

---

[22] On Mori, see Kimura, 'Masahiro Mori's Buddhist Philosophy of Robot', 75.
[23] Weizenbaum, *Computer Power and Human Reason*, 264–5.
[24] Weizenbaum, *Computer Power and Human Reason*, 273–5.

about machines, and are taken most seriously in our media landscape, are often professionally involved in machine design and marketing.[25] They can design outcomes towards what they perceive to be inevitable. For example, Ray Kurzweil argued that machines would succeed us in evolutionary history and was subsequently hired by Google in 2012. That affiliation gave him additional credibility in the marketplace of ideas and a much stronger position from which to direct the future of computing.[26] More broadly, the design and construction of technologies give concrete form to the motivations of designers, marketers, politicians, and more. Put simply, technologies are the physical manifestation of human ideas and intentions.

The clearest theoretical articulation of how physical objects make ideas and intentions durable and help them circulate in human societies comes from actor-network theory. Known primarily through the work of Bruno Latour, actor-network theory emerged out of science and technology studies as a way of understanding how knowledge is constructed. Latour argues that any society (including a scientific community) is an assemblage of human actors and the objects that provide stability for their social connections.[27] Within science, for example, the circulation of publications, grant proposals, and physical objects (e.g. borrowed equipment, lab materials, animal subjects, etc.) maintains a disciplinary field and upholds that field's intellectual commitments. Often, the physical objects have a durability that people do not. This is especially true of publications, whether printed or electronic. The exchange of students and researchers might be a form of potlatch that promotes social cohesion, but the circulation of publications can last far longer and travel much farther—after all, a single student can be present in only one location while publications can be distributed widely.[28]

---

[25] Regarding the relative popularity of apocalyptic and transhumanist authors, compare the relatively unknown and irrelevant promises of a transcendent future offered by Ettinger in *The Prospect of Immortality* and *Man into Superman* or those of Esfandiary in *Optimism One* or *Up-Wingers* to those much more prominent, though admittedly later, of Hans Moravec or Ray Kurzweil.

[26] It is worth noting that pop science books are already also a strategy in the quest for social cachet. See Geraci, 'Cultural Prestige;' *Apocalyptic AI*, 56–70.

[27] Latour, *Reassembling the Social*, 70, see also 198–9.

[28] There are other widely sharable resources for building these kinds of communities, for example, cell lines in biology. For the classic study of gift exchange and the potlatch, see Mauss, *The Gift*. I am unaware of any effort to use Mauss's work to explain scientific groups, but I expect such an effort would be a valuable one.

Consider, for example, a technological artifact well known in some parts of the world and yet in other parts of the world impenetrable to the naked eye: the saree guard. In places like India, every motorcycle or scooter is equipped with a saree guard that has two functions: to keep a passenger's saree out of the wheel well and to provide a comfortable platform for the feet of a female passenger riding sidesaddle. The presence of the saree guard indicates a host of social structures, including how people are dressed and who rides on motorcycles. In the U.S., motorcycles are most common—or at least most commonly perceived popular—among young men presenting themselves as rebellious or independent. In contrast, an Indian scientist once told me: 'my brother is the brave one in the family. He drives a car!' In India, a motorcycle is often a family vehicle because it is the most economical choice for those of limited means. The presence of a saree guard indicates all this, and it reveals expectations about gender, family structures, transportation, attire, and more. The saree guard doesn't just make people comfortable on motorcycles; it carries social attitudes that are thus visibly present in the environment.

Saree guards might present legitimate questions about how society is structured and ordered, but military, police, and corporate AI design is far more deeply implicated in such concerns. The military goals conceived at the heart of techno-apocalyptic discourse may explain why some countries reject the global push to demilitarize robotics and outlaw autonomous weapon systems.[29] Early 21st-century interest in surveillance has questions of social control at its heart. Predictive policing, which uses data analysis to suggest where crime might take place, has quietly joined the U.S. law enforcement arsenal with little to no public input.[30] The Indian government argues that authorizing a wide swath of its own agencies to monitor and decrypt citizens' computer and mobile phone activity is a pro-privacy policy.[31] Compounding this, the mandatory use of India's Aarogya Setu COVID-19 tracking app raises hackles over its long-term

---

[29] See Gayle, 'UK, US and Russia Among Those Opposing Killer Robot Ban'.

[30] Haskins, 'Dozens of Cities Have Secretly Experimented with Predictive Policing Software'.

[31] Rajagopal, 'Order on Surveillance Meant to Protect Privacy, Govt. tells SC'. Ironically, the Indian government enacted this policy *after* its own think tank promoted the importance of data privacy in the era of AI; see NITI Aayog, 'National Strategy for Artificial Intelligence', 87–8. Nick Bostrom argues that universal surveillance might also be the only way to prevent human extinction; see Bendix, 'An Oxford Philosopher Who's Inspired Elon Musk Thinks Mass Surveillance Might Be the Only Way to Save Humanity from Doom'.

potential for surveillance and an approach deemed more intrusive than necessary.[32] As Luciano Floridi points out, COVID-19 apps in general have unanticipated risks and can perversely disenfranchise citizens and may in some circumstances even exacerbate the spread of disease.[33]

In more widespread, but less emotionally charged, circumstances, Internet users have so far surrendered to the corporate presumption that each and every computer or smartphone user is a legitimate marketing target.[34] Granting corporations the right to surveil us in our Internet usage is now payment for services rendered by those companies. Partway into the 21st century, both government and corporate positions imply the inevitability, ubiquity, and acceptability of corporate surveillance through AI-enhanced big data analysis.[35] Technology producers thus choose to align intellectually and morally—because it is profitable to do so—with surveillance capitalism while obfuscating and excusing that alignment in a language of technological determinism.[36] Across the globe, sturdy resistance to surveillance has proven elusive and legally hard to establish.[37]

Surveillance and control are not, however, the only ideas deployed through AI; our computers, phones, social media outlets, and the rest of our AI-enhanced technologies bring to bear other parts of culture. Some

[32] Zargar, 'Privacy, Security Concerns as India Forces Virus-Tracing App on Millions'.
[33] Floridi, 'Mind the App'.
[34] Dijck, 'Datafication, Dataism and Dataveillance'.
[35] Zuboff, 'Big Other', 75.
[36] Ibid.
[37] The asymmetry of knowledge that leaves consumers uncertain regarding how much and which information is surveilled contributes to citizens' difficulty in resisting surveillance capitalism (Acquisti, et al., 'The Economics of Privacy'). Unfortunately, simply informing users does not necessarily lead to adequate individual choices to protect privacy (Acquisti, et al., 'Privacy and Human Behavior in the Age of Information', 514; Acquisti, et al., 'The Economics of Privacy', 448). On resistance to government surveillance, see Calo, 'Can Americans Resist Surveillance?' It is important to note, however, that surveillance comes in many forms, and corporate surveillance is perhaps the most pervasive in citizen life. Cramer suggests that the limits of protection from surveillance are contingent upon legal preservation of privacy and that shifting the argument to the matter of data discrimination would afford better protection from both government and corporate surveillance (see 'A Proposal to Adopt Data Discrimination Rather than Privacy as the Justification for Rolling Back Data Surveillance'). This approach might dovetail effectively with the idea that the ideology behind constructing machines could be shifted to think about such resistance, especially insofar as rejecting the idea that human beings *ought* to be presumed to be potential terrorists, criminals, or even seemingly harmless identities, could promote better social outcomes. Acquisti, et al. show that facial recognition can be used to acquire a considerable amount of potentially sensitive data about people using an augmented reality app—a prospect that could put enormous additional information (and the control thereby gained) in the hands of law enforcement and other forces of social control ('Face Recognition and Privacy in the Age of Augmented Reality').

of these technologies apply to our most altruistic spirits: for example, Deepak Kumar describes how information technology can benefit Indian villagers, especially women and those of low social caste.[38] Contemplating advanced AI, we must ascertain what ideas and values can be imposed on society through technological development. What ideas are carried from person to person in the public deployment of AI technologies? The religious questions occupying this chapter provide one key direction from which to think about what is and what can be possible.

The distribution of cultural ideas about technology mirrors the distribution of science and technology itself. It has often been presumed that science, technologies, and the 'right' ideas about them radiate outwards from a central axis located in Europe or, by the 20th century, Europe and North America. George Basalla famously articulated this diffusion theory of science, where the raw materials of inquiry might come from the colonial periphery but the generation of actual scientific knowledge was conducted at the centre before being diffused outwards.[39] That theoretical framework leads to the erasure of nonwestern contributions and has become unsustainable when viewed through the contemporary historical lens.[40] The early dominance of the diffusionist theory notwithstanding, more careful analysis of the spread of science and technology recognizes complexity, complementarity, and interconnectivity.

As a simple example, consider a collection of botanical samples from the Malabar Coast along with detailed notes about their geographical spread, their pollinators, their life cycles, their role in the food chain, and their medical or other human uses. Almost all aspects from such a collection would emerge out of work by native and colonial scientists in India, collectors (who may or may not be those scientists), the shared knowledge of indigenous informants, and a host of other individuals 'at the periphery'. When that collection reaches a botanist in Amsterdam, or travels from there to Moscow, it brings with it a great deal of human intellectual labor such that whatever knowledge is generated in Amsterdam

---

[38] Kumar, *Information Technology and Digital Divide*.

[39] Basalla, 'The Spread of Western Science'.

[40] For a few examples, see Geraci, *Temples of Modernity*, 39. For a robust criticism of the diffusionist claim, see Raina, *Images and Contexts*, 181–4. On circulation as opposed to diffusion, see Kapil, 'Beyond Postcolonialism . . . and Postpositivism'.

is truly generated only in collaboration with that generated in South India.[41] Moreover, the publication of data in Amsterdam should not be taken as evidence that scientists there are personally (or possibly at all) responsible for the generation of that knowledge—plausibly, such individuals are glorified (or self-glorifying) transcriptionists. There is no longer a clear geographical locus for the manufacture of knowledge and instead we see that samples, descriptions, and analyses are fundamentally transnational.

This is not to deny that science and technology—during the colonial era and continuing to the present day—remain tied to Euro-American power centres. Priorities, publication strategies, and prestige were and are connected to those power centres, and without the involvement of Euro-American educational systems and/or funding, much of 20th- and 21st-century science would be inconceivable. Despite the pervasive relevance, even dominance, of Euro-American science, however, it is easy to recognize that there is no geographically clear port for the emergence of scientific and technological knowledge. Every year, excellent research and development happens outside of Europe and North America or in collaboration between those regions and innovators all over the world. The same is true for ideas *about* science and technology. Such ideas—often packaged in literature, religion, news media, popular entertainment, or in political action—are similarly the product of complex transnational interactions.

Within the global circulation of science and technology, we see the circulation of structures of power. Some of these structures are scientific, others cross domains of human culture. In the first case, for example, science produces control through new modes of knowing. We construct knowledge of the world, shifting from one perspective to another—one paradigm to the next[42]—and in doing so, we develop new languages for those perspectives. The phlogiston theory of heat gives way to a more useful caloric theory, and so forth. Of course, the phenomena so described are very real. It's worth noting that Latour, at least, and perhaps the entire panoply of actor-network theorists in science and technology studies, were very much realists; their critics had missed the point of

---

[41] This example is only borderline fictitious; see Jain, 'A Kerala Botanist's Affair with an Unlikely 17th Century Book.'
[42] Kuhn, *The Structure of Scientific Revolutions.*

actor-network constructivism entirely.[43] The caloric theory of heat is superior to the phlogiston theory because it emerges out of a much stronger network—more instrumentation, more experiments, more publications, more explanations followed by extrapolation into new domains, more people able to make sense of it and use it, and stronger, clearer influence of the natural world (made stronger through the aforementioned techniques, apparatuses, etc.). The caloric theory, buttressed by all that support, permits much greater prediction, explanation, and hence control over our environment than the theory that preceded it.

The French philosopher Michel Foucault was at pains to demonstrate, however, there are also systems of cultural power that flow through science. Foucault argued that for any given era, the rules that govern science (and thought more broadly) should be examined holistically and that the sciences abide by the same discursive rules as contemporary disciplines. What he calls an archeological territory is a structure of possible statements about the world that includes statements from outside science itself (i.e., from literature, politics, etc.).[44]

To take an example from this book, colonial prejudice about Indians and Indian science helped shape the scientific claims of both Indians and non-Indians. The ideological prejudices that governed pre-independence India remain relevant in contemporary culture even if they have less hold than in the past. For example, when two of Boeing's 737 MAX airliners crashed in 2019 due to a software error in their management of flight data, one essay pointed towards the outsourcing of software to Indian engineers.[45] Indian engineers had nothing to do with the flight system in question, and so raising concern over their involvement shows the remnant of distrust in Indian scientific expertise and the ready way in which blame can be shifted—even without justification—towards nonwestern scientific actors. A variety of power structures are inherent in all manner of discourses, including the technological and scientific.

---

[43] Latour, *Reassembling the Social*, 88; Latour, in fact, specifically critiques those for whom science might be somehow divorced from empirical nature, see *We Have Never Been Modern*, 67. Similarly, however, he also criticizes the idea of unmediated access to nature (ibid., 82). His project was ever to show how scientific knowledge emerges through a network of practices, objects, and people—the more robust the network, the more certain the knowledge (ibid., 121).

[44] Foucault, *Archeology of Knowledge*, 183.

[45] Robison, 'Boeing's 737 Max Software Outsourced to $9-an-Hour Engineers'.

The rules of such discourse constrain what can be said within any given worldview, challenging us to consider what might be possible for an AI researcher to say about AI. Similarly, we can wonder what structures of power are inherent in AI at present and potentially in the future. If the history of science shows us that the terms and techniques of science come into being along with the development of a scientific paradigm (or perhaps Foucault's archeological territory and its attendant rules of discourse), then we are building ways of speaking about machines, especially intelligent machines, even as we build the machines. Intelligent machines are, despite some precursors, essentially novel and we are learning what it means to experience them. We learn haltingly, struggling with the inchoate responses of autocorrect systems, chatterbots, and quasi-intelligent assistants, but our lurching progress accompanies each step of technological development and deployment in society.

Already, as Selma Šabanović shows, robotics and AI are part of cultural meaning-making enterprises that construct the meaning of technology in culture. Šabanović argues that despite Japan's decades-long recognition for its 'robot culture', Japanese researchers and roboticists assiduously cultivate that image in their production of robots—it is not culturally given or automatic.[46] She explores how roboticists situate their robots within traditional histories and values of artistry, craft, and affective engagement. Importantly, this shows that even in a culture declared to be already and always welcoming of robotics, such cultural conditions are co-created as discourse, policy, and innovation.

If we are to engage in a scientific and in a social scientific study of AI, and if we are to sincerely contemplate the future deployment of AI technologies, we must ask the questions that reveal what intentions are built into the system. How is AI to be designed? What motivations should dominate design and deployment? Whose needs should be met by AI? Whose voices should be heard and in what forums—public or private? Who should have access to AI technologies?

We are further troubled in this work because AI restructures the humanities and social sciences used to understand it. As described by Foucault, those disciplines oriented towards the study of humanity organized in the 19th century in and around new scientific concepts. Their

---

[46] Šabanović, 'Inventing Japan's "Robotics Culture"'.

precariousness, he argues, hinges upon their 'epistemological config-
uration' drawn from (1) mathematical sciences, (2) the disciplines of
philology, biology, and economics, and (3) philosophical reflection.[47]
AI disrupts this triad, as it acts with more force upon other disciplines.
Foucault believed that the mathematical sciences present an uncomplicated relationship to the study of human beings (as opposed to the role
played by language, life, and labor);[48] but AI, presumably a mathematical science, poses at least as many complications to the human sciences
as do labor relations and the rest of Foucault's examples. What modes of
thought must emerge when 'humanistic' disciplines study the cultural
products of both human beings and machines which act like human
beings?

In the opening decades of the 21st century, AI could create art and
compose prose. The Next Rembrandt, a project of ING (a Dutch banking
group), Microsoft, and others, learned to reproduce Rembrandt's style
with sufficient fidelity to paint a portrait that could sit comfortably alongside the master's. Faced with such artistic accomplishment, one cannot
help but wonder what contributions AI might make to politics, theology,
and other domains. Even modest gains might appear superior to present
human decision making! Insofar as such efforts unfold, they will raise entirely new questions for the humanities and social sciences.

The nascent discipline of 'machine behaviour' takes into account
this confluence of disciplines. As Iyad Rahwan and his colleagues note,
scholars from a wide array of fields have witnessed the 'broad, unintended consequences of AI agents that can exhibit behaviors and produce downstream societal effects—both positive and negative—that are
unanticipated by their creators'.[49] They rightly point out that 'AI agents
can shape human behaviors and societal outcomes in both intended and
unintended ways'.[50] Because of this, our entire social environment is one
that demands new academic worldviews and new models for public interpretation. While we have never been insulated from the influence of
non-human actors in our cognition and behaviour, what barriers existed
heretofore are under annual revision as technology improves.

---

[47] Foucault, *The Order of Things*, 347–8.
[48] Foucault, *The Order of Things*, 351.
[49] Rahwan, et al., 'Machine Behavior', 477.
[50] Ibid., 478.

AI reconfigures our sense of what we can say about ourselves, about life, and about networks of labor; AI doubly confounds the complicated relationships the study of humanity has with these enterprises. It is hard enough to think about the meaning of life or the creation of value. When AI seems alive or AI replaces human labor, what was already hard now inches towards the impossible. Foucault claims that the human sciences are 'interlocking',[51] but when AI is considered we find ourselves at a conjunction where the study of humanity is irrevocably joined also to physical sciences. What can be produced mechanically changes how we perceive the human sciences and what constitutes the human becomes muddled. It is thus important to think through the horizons of control by which we can guide one set of organized knowledge into a future set.

Foucault deemed it impossible to discuss science without thinking about structures of power, and while his approach leaned too far away from empirical reality, the role of power and control is now central to thinking about science and technology. The transferal of technologies and ideas about them across the world cannot escape such questions. The Internet was early seen as an accelerator for technological deployment and advancement, but it simultaneously required that many countries find internal and external mechanisms for preserving local cultures and identities.[52] While recognizing the importance of Foucault's interventions, we should not permit his undue valorization of structural laws; it mystifies the world. While Foucault acknowledged the need to see how one era's rules might give way to others, he appears to maintain the ontological reality and personal agency of those rules. In Foucault's semi-mystical account, the rules act on people and society.

In powerful ways, actor-network theory rejects Foucault's mysticism while explaining how the Foucauldian rules operate. Latour argues that objects must explain the existence of power, not the other way around.[53] Latour turns against the elusive structures of Foucauldian discourse and the idea that discourse floats along by itself, suggesting instead that power is distributed by people through objects. Subsequently, both objects and

---

[51] Foucault, *The Order of Things*, 358.

[52] See Hongladarom, 'The Web of Time and the Dilemma of Globalization'.

[53] See Latour, *Reassembling the Social*, 83. Latour further rejects the idea of 'society' as a domain of reality that provides the context for the elucidation of empirical facts (*Reassembling the Social*, 4).

people act upon people. By instilling intentions and motivations in objects, they produce an architecture of power that makes some things possible by provoking people to think in particular fashion. Remember the simple example provided earlier: the saree guard and its impact on passengers and observers. 'It's the power exerted through entities that don't sleep and associations that don't break down that allow power to last longer and expand further—and, to achieve such a feat, many more materials than social compacts have to be devised'.[54]

Human decision-making constrains technology from design to deployment, and so we must explore the ideological commitments built into technology. In AI, the 'technological normativity' by which legal or social norms are embedded in the programming of algorithms diminishes public evaluation of those norms.[55] For Foucault, an archeology of knowledge meant trawling the discursive rules that make knowledge possible.[56] More concretely, however, one can construct an archeology closer to the mainstream use of that term: to uncover the artifacts that make a culture possible. An archeology of AI must account for theories of systems control (cybernetics), economic realities, computable problems (military, commercial, political, educational, etc.), uses of non-human computing machines, and a history of available technologies. It must also account for theories of social control and philosophical and religious theories of historical progress because, as we see in dreams of inevitability and technological determinism, these underwrite views of technology.

Although some technological predictions 'point toward the anti-human and apocalyptic, we can craft others that will be conducive to the long-range survival of humans and of other life-forms, biological and artificial, with whom we share the planet and ourselves'.[57] Attempts to evaluate the moral risks and rewards of AI assistants have shown this effort is complex.[58] Perhaps the idea of public reason, which Binns uses to evaluate algorithmic decision-making,[59] could be made useful in our pursuit of just AI systems. If so, we would need to find universal or near-universal

---

[54] Latour, *Reassembling the Social*, 70.
[55] D'Agostino and Durante, 'Introduction', 501; see also Binns, 'Algorithmic Accountability and Public Reason', 546–7.
[56] See, in particular, *The Archeology of Knowledge*, 129, 138–40.
[57] Hayles, *How We Became Posthuman*, 291.
[58] Danaher, 'Toward an Ethics of AI Assistants'.
[59] Binns, 'Algorithmic Accountability and Public Reason'.

moral and policy positions (e.g. equality of opportunity) and use these to undergird technological progress. In the remainder of this chapter, however, I turn towards practices and doctrines enshrined in religion that find their way into the world of AI. Visible in pop science and science fiction, these religious elements buried in the conception and construction of intelligent machines provide crucial points of leverage in the ways that future AI technologies will contribute to social order.

Religious commitments and the rendering of AI

While we must be cautious about essentializing eastern or western, Indian or American, perspectives, it remains the case that the religious perspectives dominant in the U.S. and India intersect differently with AI technologies and present distinct prospects for the future. The Apocalyptic AI movement discussed throughout this book promises a radical transformation of the world, though at possible risk to human dignity and self-sovereignty. Nascent Indian transhumanism reflects on duty—for, by, and between human beings and machines—a spiritual value that can emphasize mutual concern and an ethic of responsible care, but which can also be leveraged into social restrictions based on birth, gender, or other accidents of fate. Both U.S. and Indian visions of AI draw upon religious resources, and both are amenable to polarized interpretation: they could be used as ideological support for a better world or they could be used to more deeply entrench models of social control.

AI lends itself to visions of radical futurism, and as computers became integral to daily life, they encourage us to adopt transhumanist dreams. For obvious reasons, the impressive shifts in daily life that we experience, thanks to mobile phones, the Internet, and other computing technologies, makes it more believable that technology will permit a radical reconstruction of human bodies and minds.[60] Transhumanists in the mid-20th century may not have been particularly convincing,[61] but their movement has caught the attention of popular media, and discussions

---

[60] Geraci, 'L'Évangélisme Transhumaniste'.

[61] For histories of transhumanism that come from a variety of intellectual directions, see More, 'Philosophy of Transhumanism;' Geraci, 'There and Back Again;' Tirosh-Samuelson, 'Transhumanism as a Secular Faith' and 'Wrestling with Transhumanism;' and Alexander, *Rapture*.

of prominent transhumanists like Ray Kurzweil now circulate widely in news and entertainment.[62] Because all human beings use technology to transcend their bodily limits (e.g. eyeglasses or clothing), it would be reasonable to suggest that we are all transhumanists, and it may be that the gap between such everyday humanity and the radical pursuits of Kurzweil and avowed transhumanists is shrinking.[63] In any case, the difference between using technology to achieve a supposedly natural norm of health and using technology for enhancement is, to say the least, a fungible one.[64]

As I have argued throughout this book, the Christian heritage in the West has promoted goals of technological transcendence, and at this point it bears reiterating that this heritage also plays a role in the technological determinism at stake in descriptions of AI. In his seminal work, *The Religion of Technology*, David Noble shows how Christian expectations about divine providence and historical progress were united with the desire for salvation in European (and later American) perspectives on technology.[65] Working with one thousand years of historical documents, Noble shows that developments in technology were enthusiastically used as evidence of progress towards the return of Jesus.

The twin expectations of divine providence and technological progress were thus conjoined, forming the technological providence that—now outwardly severed from its Christian moorings—alleges the natural and inevitable unfolding of technological destiny. Many computer programmers working on artificial life simulations, videogames, and virtual environment simulations already see themselves as godlike beings in whom this destiny moves towards fulfillment.[66] A host of religious and quasi-religious theories of social transformation pervaded the rise of cyberculture,[67] and these were vigorously propped up

Singularity Cafe, founded by Jayan Prasad in 2019. The café, which is in Chennai, serves coffee with a side of Singularity theory. Prasad produced three videos on the café and what he calls 'Impact Theory', all of which

---

[62] For a variety of examples, see Goldman, 'Ray Kurzweil Says We're Going to Live Forever;' Grossman, '2045: The Year Man Becomes Immortal;' Kushner, 'When Man & Machine Merge'.

[63] On this distinction, which he identifies as a difference between 'transhumanism' and 'Transhumanism', see Hefner, 'The Animal that Aspires to be an Angel'.

[64] For a scholar who maintains that upholding this distinction is a moral and policy imperative that rejects human enhancement, see Kass, 'Letter of Transmittal to the President'.

[65] Noble, *The Religion of Technology*, see esp. 43–56.

[66] Geraci, *Virtually Sacred*, 175–6; Helmreich, *Silicon Second Nature*, 83–5, 140–1, 191, 193.

[67] Turner, *From Counterculture to Cyberculture*.

by a Christian framework even when Christian theology ceased to be a talking point.[68]

The crypto-religion of inevitability and the expectation of a transcendent future motivates the Apocalyptic AI movement. Moravec, Kurzweil, and other Apocalyptic AI authors simply swap out gods for secular guarantors such as technological determinism, evolution, or even a universal 'law of accelerating returns'.[69] Drawing on these implacable forces, they argue that our world will undergo a transformation: soon, the cosmos will be a mechanical world of digital computation. Such views inspire research and entrepreneurial agendas. In an email conjoining his commercial robot development and his views on the future, Moravec described himself to me as 'too busy making the plan happen to want to spend time talking about it'.[70] More recently, Ben Goertzel—AI entrepreneur, member of the leadership group for the world's largest transhumanist organization (Humanity+), and advocate of the intersection of Apocalyptic AI and evolution—described his work on the OpenCog non-profit enterprise as having two aspects: one of which, 'more directly exciting for transhumanists, is to use OpenCog to create a thinking machine. Granted, the aspiration is pretty big and we want it to be the next step beyond human beings'.[71] These kinds of views are widespread in AI communities.

In its apocalyptic manifestation, AI promises a radical transformation of the world—this optimism could help motivate deliberate problem-solving. Given the politically fraught realities of early 21st century life, such a transformation might be welcome indeed. Climate change-induced calamities, environmental degradation, species extinction, extreme and increasing wealth inequalities, political disenfranchisement, and polarized electorates reveal the pressing need for a new path into the future. The new world promised by Moravec, Kurzweil, and others seems to make those problems obsolete: godlike machines will clean our air and water, design sensible policies, and obviate religiously-, racially-, and ethnically-charged tribalism.

---

[68] Ali argues that these longstanding apocalyptic perspectives also continue to carry the orientalist and colonialist narratives into transhumanist engagement with AI (' "White Crisis" and/as "Existential Risk" ').

[69] See Geraci, *Apocalyptic AI*, 27–9.

[70] Moravec, email correspondence with the author, 25 June 2007.

[71] Underwood, 'The Future of Artificial Intelligence According to Ben Goertzel'. On Goertzel's apparent commitment to evolutionary pressures as a deciding factor leading to intelligent machines, see Goertzel, *The Cosmist Manifesto*, 9, 273.

Apocalyptic AI advocates mirror the two-stage apocalyptic process where an earthly, human paradise precedes an even more radical shift into a divine paradise.[72] Not only will advanced technologies supposedly eliminate pollution and clean up the environment,[73] they will provide a universal basic income that ensures we will overcome our most pressing problems in the near future. Moravec promises that we will have the leisure to enjoy a 'comfortable tribalism' and that conflict will dissolve in a 'garden of earthly delights ... reserved for the meek', which is to say humanity.[74] Ultimately, however, this initial paradise would give way to the cosmic mission of super-intelligent machines.

Isaac Asimov, probably the most influential science fiction author to tackle machine intelligence, uses the final story in his collection, *I, Robot*, to describe a world where machines (not just any machines, but The Machines) take over human decision-making. This makes the world safer and more comfortable for people but simultaneously reduces them to mere instrumentality: they live out the lives determined for them by The Machines' godlike calculation.[75] Noting through his characters that there is something both wondrous and horrible in this, Asimov prefigured the risk we face in dreams of apocalyptic transformation.[76]

Already, human beings tend to dumb themselves down in their interactions with computers;[77] should the machines become legitimately intelligent, we will be tempted to dispose of human agency altogether. Compounding this, the algorithms used to mediate all manner of digital information gathering and communication can go unexamined and ungoverned, with obvious consequences upon the human beings affected by them.[78] While problems with believing that machines know better than their human operators have plagued AI for decades, such disenfranchisement of the human intellect could produce a brave new world of human servility.[79]

---

[72] Geraci, *Apocalyptic AI*, 31–2.

[73] Kurzweil, *The Singularity is Near*, 241, 249, 251–3, 397; Drexler, *Engines of Creation*, 120–3.

[74] Moravec, *Robot*, 134, 136–7, 143.

[75] Asimov, *I, Robot*, 192.

[76] On the direct influence of theological reflection, see Geraci, 'Robots and the Sacred in Science and Science Fiction'.

[77] Lanier, *You Are Not a Gadget*, 31–3.

[78] D'Agostino and Durante, 'Introduction'.

[79] For a crucial example, see Chris Hables Gray, 'Artificial Intelligence at War'. Zarsky notes that unthinking subservience to machine decisions is easily had and could produce injustice ('Automated Prediction', 34). For a more recent engagements, see D'Agostino and Durante,

The power of AI to control human choices is relevant across a wide technological spectrum. Pattern recognition and facial recognition software, for example, can be used to identify (or misidentify) individuals and initiate protocols for exerting pressure on them. Digital advisors, such as included on mobile phones, could direct users towards one set of outcomes rather than another. Stock trading AIs already control many aspects of the marketplace while banking AIs dictate who does or does not get business loans and mortgages. As AI becomes increasingly powerful, so too does its potential for exerting control at both individual and policy levels. The difference between data modeling and control systems is a hazy one, and systems that use surveillance and feedback to control behaviour are on the horizon.[80] Human choices will dictate whether such power leads to the improvement of life on earth or promotes the mechanization of humanity.

Mirroring the ambiguous potential of American expectations of radical transformation, Indian intellectual and religious commitments to duty offer resources that could establish AIs as either beneficent allies or malicious executives of social control (note that this is not dependent on whether the machines become conscious or intelligent). Any student of Indian history or philosophy will be familiar with the many religious inflections of duty (*dharma* in Sanskrit), which extends across the cultural landscape to include obligations and observances that originate in family, gender, caste, lifecycle, and religious participation. For some Indians, adherence to one's duty is a crucial component to one's personal fulfillment and salvation. The ancient sage Manu, for example, clearly details the expected duties contingent upon a man's stage of life (student, householder, forest-dweller, or ascetic) and caste (more properly *varna*), and those obligatory for women.[81] While actual custom may have—indeed, often did—vary significantly from Manu's elucidation,[82] the centrality of

---

'Introduction', 502; Pagallo, 'Algo-Rhythms and the Beat of the Legal Drum', 508–11. See also Dennett, 'The Singularity—An Urban Legend?'

[80] Van Otterlo, 'Automated Experimentation in Walden 3.0;' Palmås, 'Predicting What You'll Do Tomorrow'.

[81] Manu, 'Laws of Manu'.

[82] For example, on the complexity of caste (both *varna* and *jati*) in Indian history, including its fluidity in the pre-British period, see Chatterjee, *The Nation and Its Fragments*, 119, 166; Klostermaier, *A Survey of Hinduism*, 43; Banerjee-Dube, *A History of Modern India*, 143; Srinivas, *Social Change in Modern India*, 95–124; Perera, *Debating the Ancient and Present*, 31; Nath, 'From "Brahmanism" to "Hinduism"'; Singer, *When a Great Tradition Modernizes*,

duty has never wavered in Indian religious thought. In Indian epics, tales, and ritual injunctions, duty dictates the behaviour of gods, demons, and heroes alike.

Perhaps the first non-Indian to advocate the clear integration of Hindu duty to science was the proto-transhumanist J.B.S. Haldane. Haldane suggested that Indian scientists could become global leaders by integrating a Gandhian principle of non-violence into science.[83] For example, he believed that most Indian readers 'will agree that one of the contributions which India should make to world ethics is the inclusion of duty to animals within the sphere of human duties' and that this ethical stance can and should be applied within biology.[84] Haldane's biological studies in India were themselves a significant challenge to the centre-periphery model promoted by Basalla and his acquisition of new methods and modes of reasoning in India unquestionably affected his science and the global study of genetics.[85]

Duty holds a privileged place in Indian science, thanks especially to the legacy of Jawaharlal Nehru, India's first Prime Minister.[86] Nehru believed that science is crucial for 'social development, of the ceaseless adventure of man'[87] and that scientists have a special obligation in this: they are to 'develop more of what in India we consider the Brahminic spirit of service'.[88] The spirit of service, once integrated into science, ensures that science would confirm its real value, that it 'widens the spirit of man and thereby betters humanity at large'.[89] Following Nehru, many scientists

345, 354; Hansen, *The Saffron Wave*, 35; Misra, *Vishnu's Crowded Temple*, 33–42; Gottschalk, *Religion, Science, and Empire*, 205–7; Nair, *The Promise of the Metropolis*, 53; Sathaye, *Crossing the Lines of Caste*. Some scholars go so far as to deny that caste had any significant role in India prior to the arrival of the British (Balagangadhara, *Reconceptualizing India Studies*, 4, 55, 194, 235, 237; Gupta, *The Caged Phoenix*, 160).

[83] Haldane, 'Some Reflections on Non-Violence', 139.
[84] Haldane, 'A Biologist Looks at India', 276. A brief summary of Haldane's work in India can be found in Dronamraju, 'J.B.S. Haldane's Last Years'. Haldane's rejection of animal cruelty was complicated as he does not oppose any experiment that he would willingly perform on himself (see 'Some Reflections on Non-Violence', 135–6), but he certainly rejects much of the violence he sees in science; see also Shiv Visvanathan's critique of modern science in *A Carnival for Science*.
[85] McOuat, 'J.B.S. Haldane's Passage to India'.
[86] For a more extensive evaluation of Nehru's integration of nationalism and science, see Geraci, *Temples of Modernity*, 52–6.
[87] Nehru, *Discovery of India*, 19.
[88] Nehru, 'The Need For a Spirit of Service', 44.
[89] Nehru, 'The Spirit of Science', 79.

and engineers in India believe that scientists have a clear duty to the society in which they operate, arguing that scientific work should make life
better for their fellow citizens.[90] The rhetorical power of duty can be sufficiently strong to provide the entire justification for Indian engagement
with futurist technology.[91]

As Indians contribute to the global development of AI, it is thus no
surprise that the religious value of duty is relevant to the ways in which
Indians reflect on the advancing technologies. Early visions of AI in
Indian science fiction were heavily dependent upon Isaac Asimov,[92] but
more recent contributions build on that while departing in significant
ways. India's rich religious and cultural history means that duty is not the
nation's only possible contribution to popular or technical considerations
of AI, but theorizing duty could be a significant contribution of India to
global AI design and, indeed, it is already present in Indian cultural references to the technology.

A shared sense of duty is the key to human beings and machines flourishing in Indian science fiction, which could affect the technical contributions of Indian scientists and engineers. Shared duty is what brings
humanity and AI together across the gulf of meaning that has often separated the two. AI pioneer Joseph Weizenbaum, for example, sought to
articulate a difference between human beings and machines: 'man, in
order to become whole, must be forever an explorer of both his inner and
his outer realities. His life is full of risks, but the risks he has the courage to
accept, because, like the explorer, he learns to trust his own capacities to
ensure, to overcome. What could it mean to speak of risk, courage, trust,
endurance, and overcoming when one speaks of machines?'[93]

While Indian scientists may sometimes miss the value of their own
cultural resources in the future of AI,[94] Indian science fiction offers a
counter to Weizenbaum's doubt. In Sujoy Ghosh's film *Anukul* (adapted
from a 1970s era story by Satyajit Ray), studying the *Bhagavad-Gita* leads

---

[90] Geraci, *Temples of Modernity*, 68–74; Sekhsaria, *Instrumental Lives*, 20–2.
[91] This occurs in science fiction; see, for example, Thakur, *123 Tomorrows*, Kindle location
1014, 1093.
[92] For example, see the contributions to Phondke, *It Happened Tomorrow*.
[93] Weizenbaum, *Computer Power and Human Reason*, 280.
[94] For example, in one of the few instances of an Indian publication wrestling with Apocalyptic
AI claims, Govind Bhattacharjee relies entirely on western sources as he considers the ramifications of intelligent AI and human-machine hybrids; see 'Age of Man-Machine Hybrids'.

a Sanskrit teacher and his robot servant to realize their shared obligations towards one another and, indeed, the robot takes the very kinds of risks that Weizenbaum suggests are beyond the concerns of machines.[95]

Conceiving of AI as a tool for expressing social duty offers an opportunity for the technology to enrich human lives. Indian scientists and engineers frequently express sentiments that align them with Nehru's vision of social service.[96] One software engineer, for example, argues that social problems are distressing not only to him personally but also to 'most of the technologists here. So I'd say there's definitely a lot of need for technologists like us to collaborate with others and try and find solutions for the common man'.[97] The more powerfully scientists and engineers adopt an ethos of duty the more likely that ethos could be incorporated into AI developments. Another software engineer in India argues, 'technology has been an enabler for huge, scalable social change'.[98] Reinforcing structures of reciprocity and concern through AI could go far beyond the current capacity of information transparency and effective communications that have contributed to the daily lives of Indians.

Interest in *swaraj* (self-rule) could accompany commitments to *dharma*. Care and concern for others lies at the root of India's *swaraj* movement in science and science fiction, offering the promise of inculcating the religious and political ideal of *swaraj* in computing technologies. The most famous advocate of *swaraj* was, of course, Mohandas K. Gandhi, for whom economic, political, religious, and other social systems ought to maximize individual self-mastery and control.[99] Gandhi, notes Shiv Visvanathan, employed *swaraj* and *swadeshi* (self-sufficiency) in ways that are clear precursors to today's alternative energies and alternative epistemologies.[100]

But while Gandhi saw *swaraj* and global economic power networks to be mutually exclusive (i.e. true *swaraj* meant disassociating with the

---

[95] Ghosh, *Anukul*. Similar motivations appear elsewhere in Indian science fiction, such as when a consideration of justice and duty lead an AI to change course after planning to execute revenge against humanity in the story 'Sita's Descent' by Indrapramit Das.

[96] See Geraci, *Temples of Modernity*, 69–74.

[97] Quoted in Geraci, *Temples of Modernity*, 73.

[98] Quoted in Geraci, *Temples of Modernity*, 73.

[99] Gandhi, *Indian Home Rule*, 45–7, 65; Geraci, *Temples of Modernity*, 46–8; Vajpeyi, *Righteous Republic*, 58.

[100] Visvanathan, 'Forward', Kindle location 38.

networks of manufacturing, trade, and organization that stretched around the nation and world), many Indian thinkers believe that those very networks can be harnessed on behalf of individuals to alleviate social problems. Carl Malamud describes Sam Pitroda, for example, as a 'technology optimist' for whom technology ought to solve disease, communication, and environmental degradation.[101] Deeply interested by his youthful adherence to Gandhi and Gandhian advocacy, Pitroda claims that these problems and more can be solved with a focus on 'human needs' and 'non-violence'.[102] Without explicitly formulating a theory of *swaraj*, many Indian approaches to digital technology already work towards it by enhancing the flow of information to villages and the power of villagers to respond dynamically to the economy.[103]

Science fiction likewise summons the spirit of *swaraj* to technology. Vinayak Varma speaks of 'Swarajians' in his short story 'Omni8'.[104] Aside from whatever motives the Swarajians might have, Varma's lighthearted story optimistically portrays the possibility that robotics and nanoengineering might restore a blighted landscape, bringing life to a toxic world.[105] Vandana Singh describes a combination of traditional and digital technologies that allow local communities to flourish after climate change has ravaged the world.[106] The smart cities of Singh's speculative future offer a particular glimpse of *swaraj* for the dispossessed and marginalized.

On the basis of *swaraj*, Indian advocates have begun building a new scientific ethos that foregrounds 'the values of sustainability, plurality and justice'.[107] In *Knowledge Swaraj: An Indian Manifesto on Science and Technology*, participants of the Knowledge in Civil Society Forum declare that their work 'is not just for India, but a modest offering from India to the world'.[108] Based on Gandhian values, they argue for public trusteeship over knowledge,[109] a practice that could increase collective

---

[101] Malamud, 'Note on Visit to Sabarmati Ashram', 21.
[102] Pitroda, 'Right to Information, Right to Knowledge', 63; Pitroda, 'Access to Knowledge in America and India', 34.
[103] See Kumar, *Information Technology and Social Change*.
[104] Varma, 'Omni8', 68.
[105] Varma, 'Omni8', 89.
[106] Singh, 'Reunion'.
[107] Knowledge in Civil Society, *Knowledge Swaraj*, 5.
[108] Ibid., 6.
[109] Ibid., *passim*.

participation in knowledge generation and in the social influence of science and technology.

While the authors of *Knowledge Swaraj* focus on concerns shared by only some nations (e.g. the agricultural practices that disadvantage and even disable Indian farmers), they recognize that their conceptual framework applies to the entire range of modern technologies (e.g. nanotech). Naturally, *swaraj* is a powerful starting point for a critique of surveillance and control technologies. If we take seriously that individuals have duty to one another, and that part of this duty is to promote individual self-mastery, then we can conclude that political, economic, and military surveillance must be subject to strict limitations. Giving AIs the keys to our surveillance regimes would fly in the face of *Knowledge Swaraj* and its advocacy of public trusteeship. The same is true for other forms of social control through political or economic mechanisms. Indian perspectives on duty and *swaraj*, inflected as they are by Indian religious and cultural traditions, offer powerful tools for reinterpreting and redirecting technology towards humane ends.

The history of caste-based conflict, however, raises the spectre of abusive control through AI. Malign interpretations of duty are obvious and easy to come by, such as the thought that we might reaffirm undesirable duties and lives for human beings or intelligent machines. Manu, the lawgiver, prohibited social movement by declaring that 'it is better to discharge one's own (appointed) duty incompletely than to perform completely that of another; for he who lives according to the law of another (caste) is instantly excluded from his own'.[110] Could the valorization of duty, then, justify social conditioning and the subjugation of individuals? At least in theory, emphasizing individual duty can promote enforcing specific duties on individuals. That is, of course, not only the perceived danger of India's traditional caste system (especially as codified under British rule) but also a clear and present danger in visions of the future for all nations: from Aldous Huxley to contemporary Indian fiction, the threat of forced caste identity haunts the world to come.[111]

The presence of religious influence upon AI is not inherently good or bad, but the fact of such influence suggests that religious motivations

---

[110] Manu, *Laws of Manu*, 184.
[111] For examples, see Huxley, *Brave New World*; Chabria, *Generation 14*.

might become morally charged technological motivations. One might ask whether the transformation of the world alleged in American interpretations or the structures of duty inherent in Indian could be used to undermine the ethos of surveillance and domination that pervade political, commercial, and military regimes of control. Norbert Wiener, considered the founder of cybernetics, argued that 'a goal-seeking mechanism will not necessarily seek *our* goals unless we design it for that purpose ... The penalties for errors of foresight, great as they are now, will be enormously increased as automatization comes into its full use'.[112]

The early 21st century revealed AI bots' influence on political practice, increasing concern over Internet privacy, and a host of other anxieties.[113] If we are to alleviate our doubts about AI, we must uncover the hidden movements in technological design and direct these towards human ends. Both apocalypticism and *dharma/swaraj* have potentially dangerous applications. But understood and judiciously directed, they could deter inhuman social control and provide valuable direction towards technological development.

Conclusion

Whether or not AI achieves or exceeds human equivalence, it will have real motivations. Either there will be sentient machines, which by definition would have motivations, or there will be non-sentient machines that incorporate human motivations built into the systems in the same way that all technologies reflect human intention. As a consequence, one must question how to think about machines, how to think about different nations' contributions to our technological future, and how to apply leverage to the recursive process of technological design.[114]

---

[112] Wiener, *God & Golem, Inc.*, 63, emphasis original. In keeping with this book's themes, it is worth noting that cybernetics as a field of study directly impacted the development of transhumanist thought. See, for example, Huxley, 'The Future of Man', 8.

[113] On AI and transhumanist anxiety, though not the political imbroglios referenced above, see Singler, 'It's the End of the World as We Know It'.

[114] Surprisingly, McGrath and Gupta suggest that a self-aware machine's values should be 'hard-wired (literally or metaphorically) so that neither they nor their priority can be changed' ('Writing a Moral Code', 2). Given the seeming impossibility of establishing universalizable values it seems like a self-aware machine should be able to learn and grow. Subsequently in their essay, McGrath and Gupta appear to recognize this (p.12).

How we think about AI has consequences for the integration of AI into political, economic, and personal life. We can think of machines as tools for mind uploading and departure from the human condition, for building a different kind of society, for ameliorating human misery, for tracking human behaviour and encouraging specific choices, for anticipating human choices and reacting to them, and more. These are not mutually exclusive options, and most of them play out differently depending upon the context of their deployment: commercial, civic, education, health, etc. What we hope to accomplish influences what relationships we build into the machines.

In any given society, we must attend to the hierarchical ordering of moral priority and this demands attention to religious values. As McGrath and Gupta instruct, 'the inevitability of conflict arising between ethical and moral values that an individual or a society subscribes to makes it appropriate to explore solutions in a manner than brings Jesus, Isaac Asimov, and others into conversation with one another'.[115] For these authors, a conversation between religion and science fiction (a task frequently undertaken in this book) is vital to building ethical AI systems. The question of moral priorities that McGrath and Gupta raise is one that must be resolved within each culture and is all the more pressing when we consider the interconnections among world cultures.

Different cultural perspectives could result in very different technological approaches to those areas of human life, and thus the development of different technologies. From where we stand, there is no one technological outcome, no one predetermined future. A global perspective on AI can promote better technological outcomes. This is not to champion essentialist and orientalist perspectives: I do not suggest that India is spiritual while the West is scientific or that India is good while the West is bad. Since Said's criticism, it has been clear that such orientalist stereotypes mislead in the extreme.[116] S.N. Balagangadhara goes further, claiming that western analysis is so biased in its orientalism that it reveals nothing at all about its ostensible subject matter and reflects upon only the culture of origin.[117] While that is too strong a claim, it is nevertheless important

---

[115] McGrath and Gupta, 'Writing a Moral Code', 6.
[116] Said, *Orientalism*.
[117] Balagangadhara, *Reconceptualizing India Studies*, 41.

to separate the search for culture-specific perspectives from the manufactured differences of orientalist fantasies.

Obviously, the apocalyptic perspectives visible in the U.S. are decidedly religious, and they have the potential to benefit or endanger humanity. Their optimism can motivate us; their dismissal of biological life could lead undermine humanity. In addition, Indian values of duty mirror the double-edged nature of apocalypticism: they offer a glimpse into a brighter future of mutual concern and individual independence (particularly when *swaraj* motivates thinking about *dharma*) but also a terrifying dystopia of subjugation. Neither the American nor the Indian approach is 'the good' or 'the bad' way of thinking about AI, but they are different and they do pose different prospects for the future.

The conversations we must have to orient the future of AI include constituents from scientific, humanistic, and social scientific disciplines, and so there is pressing need for finding common ground (in method or fact) and ways of resolving intellectual differences. Nithin Nagaraj of the National Institute of Advanced Studies in Bangalore has insightfully suggested that such work resembles interfaith dialogue.[118] Given the differences in cultures, both geographic and disciplinary, best practices from interfaith discussion could be enormously valuable as we pursue global benefits.

Furthermore, the use of public reason can help us differentiate between the positive and negative directions by which religious cultures inform technoscientific progress. When judging the transformative aspirations of American technology, for example, we might ask whether these transformations benefit all people equally. A similar logic might be applied to the conceptions of duty that emerge from Indian perspectives. Public reason could provide key assistance as we look to religion for conceptual tools—allowing us to benefit from the experience of religious practitioners without requiring non-practitioners be subjugated to specifically religious logic. In such a collaboration of multiple cultures, of multiple religious perspectives, and of a search for public reason, we can promote technological progress that counts equally as human progress.

What we design machines to do, either inadvertently or purposefully, will come back to structure our society. In a different technological

[118] Nagaraj, 'AI'.

environment, with no awareness of the coming computer revolution, Haldane argued that 'scientific research would be the better for adopting a few Gandhian principles, one of which is to regard machines as made to serve men, and never to think of men as made to serve machines'.[119] The outstanding progress in AI makes such advice all the more relevant—though it will require some revision should machines ever approach humanity in their cognitive and emotional powers. A conscious awareness of cultural interests and practices might provide leverage for better outcomes as we regard near-term and long-term AI development. Given the significance and power of AI, this seems vital for human flourishing.

[119] Haldane, 'Simplifying Astronomy', 194.

5

(Re)designing the Future

Introduction

Despite allegations that religion and science exist separately or that science operates in cultural vacuum free of society, it should be more than apparent that there exist a variety of interesting points of intersection among religion, science, and technology, including artificial intelligence (AI).[1] The rise of Apocalyptic AI and its global diffusion and transmutations already impact how we see AI on corporate, public, and policy levels. A broader appreciation for how human values contribute to our perception of AI will open the door to better consideration of our technological future.

Although I am alert to the fact that global power distributions pose difficulties for creating parity among nations, we must work to expand our expectations about how AI technologies get conceived, culturally situated, and deployed in our lives. It has been suggested by one of my critics that I am perhaps blithely unaware of how difficult it would be to transform western ideologies through Indian (or other) contributions.[2] While mere words to the contrary cannot by themselves undermine such an accusation, my own intellectual positions have shifted thanks to my engagements with Indian scientists and social scientists, and I take this as evidence that other people can likewise expand their expectations and interests. Deeds must follow words, but nevertheless it matters what claims we stake. This book argues that human flourishing depends upon a global perspective and that our visions of AI should accommodate multiple cultures rather than be determined by a new form of mental colonization where one set of values dominates all horizons.

---

[1] I elaborate on the complexity of religion–science–technology interactions in Geraci, 'A Hydra-logical Approach'.

[2] See Ali, 'Review: *Temples of Modernity*'.

Futures of Artificial Intelligence. Robert M Geraci, Oxford University Press. © Oxford University Press 2022.
DOI: 10.1093/oso/9788194831679.003.0006

A naïve accounting

Throughout this book, two oversimplifications permitted me to remain on task and allow us to understand the flow and transformation of
transhumanist aspirations in AI. First, I spoke of Jewish and Christian
apocalyptic traditions with an emphasis upon their unifying themes rather than the individual distinctions across time, space, and community.
Second, I used a variety of contentious terms without much elaboration
of the complexities in their scholarly usage. These include terms such as
western, nonwestern, Euro-American, and Hinduism. Even the word religion has questionable political implications and intellectual challenges.
Although I noted in the introduction that I would take certain liberties,
I now briefly explain these choices and offer whatever apology might be
relevant for them.

Regarding my references to Jewish and Christian apocalypticism,
I make few distinctions here because my principal interest lay in exploring
what was inherited by western roboticists and AI thinkers. Within both
Judaism and Christianity, one finds unique apocalyptic movements and
texts, but this book is not one to unpack all the textual and social differences present in these traditions. I do not deny the historical and sociological distinctions at play but focus on the larger worldview indicated
by the texts and histories of apocalypticism. Different apocalyptic movements have had correspondingly different effects on their societies and on
history. There are specific ways in which apocalypticism leads one community to behave in response to historical conditions. But reference to
such would have led us far afield from Apocalyptic AI. I kept Jewish and
Christian apocalyptic traditions together as a way of conceptualizing the
cultural reference points available in the construction of Apocalyptic AI.

Second, much of this book accepts that there is *meaningful, though not
perfect,* analytical clarity in referring to things such as Euro-American or
western culture and *meaningful, though not perfect,* analytical clarity in
words such as religion and Hinduism. This is not because these categories
are uncomplicated or because any monolithic version of western identity
or Indian Hinduism has ever existed. Nor is it because I failed to notice
other religious traditions in India or the fact that whatever might constitute western or Indian culture gets defined in relation to the other. Global
history and culture are rife with mixtures and intricate interconnections.

Europe engaged in trade and cultural contact with India before even Alexander's empire; this contact permitted mutual growth and was sustained through persistent trade networks. But the distinction between western and nonwestern is still relevant because human beings use these words in shared, meaningful ways: the words are complex and sometimes fraught, but they remain productive for communication. They certainly hide all manner of important concerns, including matters of genuine justice for oppressed people. But within the context of this book, one can meaningfully suggest that Euro-American perspectives on technology can be articulated in a coherent fashion and the same is true of India. I have not attempted a comprehensive account of such perspectives but hope that I have provided relevant insight into the specific visions of technology by focusing primarily on apocalyptic and transhumanist views of AI.

Phrases and concepts have histories and politics of their own, which can be obscured by uncritical usage. Western, nonwestern, Euro-American, Asian ... the indiscriminate use of these categories has laid troubling terrain. Other important terms, such as Hinduism and even religion, are similarly open to contest. The category of religion, for example, was defined and weaponized during colonial expansion: recognition of indigenous traditions as being religions or not was used as an excuse for conquering Africans and Asians.[3] In India, recognition that the term religion emerges out of Christian culture and does not easily map onto Indian terms or concepts for apparently or allegedly similar phenomena (e.g. the existence of gods or ritual practices) goes back at least as far as Sri Aurobindo in the early 20th century.[4] More recently, a host of scholarly work has unveiled the hidden politics and problems in terms such as religion and Hinduism.[5] Recognizing the political and

---

[3] See Chidester, *Savage Systems*; Chidester, *Empire of Religion*.

[4] Ghosh, 'The Renaissance in India', 65.

[5] For examples, see Asad, *Genealogies of Religion*, 29; McCutcheon, *Critics Not Caretakers*, 10; Appiah, *There Is No Such Thing as Western Civilization*. For opposing perspectives, see Laine, 'Mind and Mood in the Study of Religion;' Doniger, *On Hinduism*, 9; Bhargava, 'An Ancient Indian Secular Age?' 189–90. One can witness the importance of such arguments by seeing the debate between Sutton ('So-Called Caste') and de Roover ('A New School in the Study of India?'), which engages whether S.N. Balagangadhara's account of Indian culture and its extrapolations by his associates at the University of Ghent are correct in their criticism of terms like Hinduism and caste. I am sympathetic to the point that Balagangadhara, the Ghent School, and others make with regard to the conceptual imperfections of terms associated with Hinduism (what de Roover calls 'cognitive inadequacies' on p. 280 of his essay), but they go too far when they seek to discard

intellectual difficulty of terms such as religion, western, nonwestern, etc.—and the political nature of their employment—is critically important but must not be all-consuming.

We cannot simply discard words that have meanings for those who use them. In India, for example, one doesn't have to wait long before hearing someone use the term 'Hindu' to refer to his or her beliefs and practices. To strip that choice from the speaker is unhelpful.[6] However complex the history of Hinduism might be, there is no question that Hinduism is not Christianity (though some traditions do find ways to intertwine). So, while Hinduism is a term that originated outside India's local communities and religion is a term culled from colonial encounters and while there never were clean breaks between western and nonwestern economies, cultures, or peoples ... nevertheless, we take what value we can from these terms.

Hence, a touch of naiveté underlies this book's approach. We permit the oversimplifications of Jewish and Christian apocalyptic traditions and the existence of historically and culturally meaningful distinctions between Euro-American or western culture and Indian culture even if the meanings of the latter terms are more complicated than the way they have been expressed in narrow, simplified, and all-too-often colonial usage. We acknowledge that conceptual categories such as religion and Hinduism are embedded within linguistic, philosophical, and experiential structures that limit what those words can mean and, importantly,

---

terminology altogether as some kind of paradigm shift. If they recognize, as de Roover does (p. 280), that both their academic disputants as well as a near universal array of Indian actors all use the term Hinduism, for example, what sense is there in trying to allege that Hinduism doesn't correspond to anything at all in reality? What alternative is there, even given the fungibility and flexibility of the present term? In 'A Blindness about India', Rajan Gurukkal points to the general absence of any progressive research arc in Balagangadhara's work: there is a critique, but no alternative (p. 13). Furthermore, despite de Roover's claims (p. 274), it is flatly untrue that the Ghent School cannot be aligned with Hindu nationalist movements simply because the Ghent School academics claim to deny the existence of Hinduism as anything other than a western obfuscation of reality: as Gurukkal points out, Balagangadhara attended the World Hindu Foundation's 2014 Congress that had 'the avowed objective of forging Hindu solidarity at the global level for rebuilding the spiritual and material heritage of India' (see Gurukkal, 'A Blindness about India'). Such direct political, participatory alignment indicates a clear engagement with Hindu nationalism whether or not Balagangadhara thinks the word Hindu has any meaning.

[6] More than unhelpful, taking away the local usage actually aligns with colonial forms of political and discursive hegemony that are supposedly disrupted by scholarly inquiry into the politics of terminology. It repeats the process of taking people's voices away from them. Indeed, while I am not a fan of V.D. Savarkar's political philosophy or his actual definition of the word Hindu, he speaks eloquently in his desire to claim the terms 'Hindu' and 'Hinduism' in his famous book, *Hindutva* (pp. 4–10).

might exclude other important meanings. In drawing some concluding insights from this book's chapters, I apologize for what I deem necessary manglings of precision. In order to be precise about one thing (Apocalyptic AI and the global crossings of transhumanism), I have sacrificed precision elsewhere. I believe some degree of coherence has nevertheless emerged and I now turn toward the outcomes.

Irretrievable truths of fiction

Library separations of fiction and nonfiction do not produce mutually exclusive domains: both make claims about the world and such claims can be legitimately truthful but also imaginative. Fictional accounts often make truthful statements about the present, which is why I have relied upon them in this book; but it is also the case that fictional accounts can transform the future and align the world with their intentions. I do not limit this perspective to the science fiction narratives which I have cited throughout. Although found in the nonfiction section of libraries, the prophecies of the iron horsemen are fictions about the future, fictions with which their authors hope to shape reality. Such narratives are profoundly important: the narration of science in fiction or nonfiction impacts culture by giving voice to certain people and removing others from the cultural discussion.[7] Through both the technologies they build and the words they publish, the iron horsemen contribute either deliberately or not to a worldview in which the concerns of transcendent posthumans and machines are more important than the concerns of human beings. Their advocacy could well bring about the world they seek.

In the remarkable story, 'Tlön, Uqbar, Orbis Tertius', Jorge Luis Borges describes a scenario where crafting a narrative is simultaneously the crafting of a world. In the story, which is set on Earth, a group of companions find a rare book with a fabricated entry about the country of Tlön. They conspire to invent more fictions about the country, and eventually its entire planet. Their publication of the First Encyclopedia of Tlön leads to an 'intrusion of this fantastic world into the world of reality'.[8] As more

---

[7] Subramaniam, *Holy Science*, 35.
[8] Borges, 'Tlön, Uqbar, Orbis Tertius', 16.

people read and learn about Tlön and its fictional culture, they transform their own world to match it. Near the end of his tale, Borges notes that 'the contact and habit of Tlön have disintegrated this world ... A scattered dynasty of solitary men has changed the face of the world'.[9]

Similarly, advocates of Apocalyptic AI describe a near-future world where machine intelligence replaces biological intelligence, where immortal minds built from zeros and ones establish a new paradigm for life on Earth and across the cosmos. The published labors of Moravec, Warwick, and Kurzweil operate on an evangelical level: they work on readers to produce an aura of inevitability that readers will then help bring about. J.B.S. Haldane, whose scientific futurism had such a dramatic effect on the rise of transhumanism, once wrote that 'I have sketched my own utopia, or as some readers may think, my own private hell. My excuse must be that the description of utopias has influenced the course of history'.[10] Clearly, he recognized the power Borges notes: we can tell stories that remake the world in their image. The iron horsemen of the apocalypse show their commitment to this task but so too do intellectual allies in India like D.K. Wadhawan and Govind Bhattacharjee.[11]

This future world is already under construction, thanks to the joined hands of global capital and policymaking. Whether human equivalent AI or mind uploading or other transhumanist desiderata are actually possible is less at stake than the choices being made about how to orient our technologies and our social outcomes. The obvious impact that these technologies have for medical therapies ensures that we will relentlessly pursue advancements in these areas. Already, privately funded initiatives investigate brain–computer interfaces (BCI) to improve Parkinson's, Alzheimer's, and many other medical conditions; but the scientists involved are quite well aware of the potential to radically enhance humanity along the lines predicted by Warwick and others.[12]

One public figure who deserves recognition as perhaps the leading horseman of the 21st century is Elon Musk. Although Musk's cultural

---

[9] Borges, 'Tlön, Uqbar, Orbis Tertius', 18.

[10] Haldane, 'Evolutionary Possibilities in the Next Ten Thousand Years', 361.

[11] It may well be that many religious leaders envy the clarity of scientists' (admittedly often dubious) futurism: scholar and theologian Martin Marty, for example, aspires toward a 'usable' view of the future (Marty, 'Still Searching for a Usable Future').

[12] See Velasquez-Manoff, 'The Brain Implants that Could Change Humanity'.

perspectives are the outcome of work by the horsemen who preceded him, he has leveraged his expansive technical interests, wealth, and considerable intellectual curiosity to produce a technological empire that aggressively pursues a transhuman future. In particular, his companies SpaceX, which plans to move human beings to Mars, and Neuralink, which pursues integrated BCI to prevent humanity's obsolescence in a world of transcendent AI, both drive toward the expansion of the human experience. Musk optimistically sees the potential for technology to create global sustainability—as long as people remain committed to the task—and also the potential for developing space colonies to protect the human species from accident.[13] He hopes that neuralink, which will begin as a therapeutic technology to solve brain and nerve trauma, will eventually lead to human transcendence. Musk reports that he 'created the company specifically to address the AI symbiosis problem, which I think is an existential threat. I mean, the reason I'm doing these things, at least aspirationally, is to maximize the probability that the future is good'.[14]

In a prior chapter, I mentioned how Musk became one of the key technologists warning about the possibility of an AI takeover, but it is equally important to note that his response to this involves clear investment in human transcendence. Just as Musk's own quest for transcendence includes his intent to live out his days on Mars, his cyborg dreams include BCI technology that will supposedly enhance human minds and resolve the complications of an advanced AI future.[15] Musk's faith is strong enough that he has staked the money to bring his cyborg future to fruition.[16] Unlike the Book of Revelation's numerically limited horsemen, the iron apocalypse—the AI apocalypse—is one that admits of new champions who carry both the technology and the movement into the future. Giulio Prisco, founder of the Turing Church transhumanist community, notes that 'if [the] Turing Church has saints, Elon is certainly one of them'.[17] One might believe that as some of the apocalyptic promises

---

[13] See his interview with Swisher, 'Elon Musk'.

[14] Ibid.

[15] Becque, 'Elon Musk Wants to Die on Mars;' Heilweil, 'Elon Musk is One Step Closer to Connecting a Computer to Your Brain;' Rapier, 'If You Can't Beth Them Join Them;' Cookson and McGee, 'Cyborgs'.

[16] Etherington, 'Elon Musk's Neuralink looks to Begin Outfitting Human Brains'.

[17] Prisco, 'Pigs in Cyberspace'. It's worth noting that Prisco does not mean that he admires Musk's ethics or personality, simply that he sees Musk bringing about a redemptive future.

become stale so would the movement, but the rise of new champions like Musk proves that this hope pervades early 21st-century views of AI and remains a vital part of the conversation about humanity and its technological pursuits.

In his essay outlining necessary steps toward the ethics of robotics and AI, John Tasioulas notes that it can be 'hard to disentangle realistic future scenarios from mere science fantasy. In light of this, our ethical thinking ... must be sensitive to the timeframe in question, sometimes addressing matters of immediate concern, other times anticipating future developments. A persistent danger is that we are distracted by potential developments that will arise, at best, in the very remote future, while neglecting pressing concerns in the here and now'.[18]

This book has occasionally woven pressing ethical conundrums (e.g. surveillance) into a narrative about what Tasioulas, but not Kurzweil or Musk, would consider distant ones (e.g. superintelligence). This is for two reasons: (1) the Apocalyptic AI narrative is culturally authoritative regardless of its near-term probability and (2) the immediate concerns of surveillance, unemployment, and human endangerment (through weaponized drones or self-driving cars rather than through Terminator scenarios) are actually connected to the Apocalyptic AI mythos. Insofar as an aura of inevitability, moral necessity, and transformative outcomes infuses the public engagement with AI, those features will be broadly applicable. In essence, the popularity of Apocalyptic AI has direct relevance for other ethical conversations about AI even if its projections prove faulty.[19]

The public commons of religion, science, and technology

This book shows that the academic study of religion, science, and technology has public relevance that depends upon its engagement with people's lives and its capacity to offer insight that matters at policy levels.

---

[18] Tasioulas, 'First Steps Towards an Ethics of Robots and Artificial Intelligence', 54.
[19] This is generally true of narratives about AI, including those in science fiction and elsewhere. For a parallel argument, see Cave, et al., 'Introduction', 7. Indeed, the narrative visions of AI are already under contest in toys and games, which are shaped by cultural influences but also help drive cultural change (see Giddings, 'Toying with the Singularity').

While writing this book, I had the opportunity to visit tech entrepreneurs in Bangalore to discuss many of its central themes. Some were well familiar with transhumanist claims, but others were much less so. All were at least interested in a discussion about the ethics of AI. Academic work should include outreach to business and policy communities in order that those groups can conduct their work to a standard of excellence that they, themselves, desire. To make that possible, academic work must draw on methods that ground research in the actual work and lives of scientists and engineers. It is not enough to discuss religious beliefs or scientific theories; we must study the actual practices of people to produce meaningful dialogue.

This new paradigm in the study of religion, science, and technology evaluates the direct impacts of religion and science in the lives of individuals and does so with a clear awareness of postcolonial power relations in which international scholarship takes place. To build the new paradigm, scholars moved toward the collection and interpretation of empirical data (most especially through ethnographies based on participant-observations and interviews) rather than the single-minded attention to texts, doctrines, and scientific theories which dominated early research.[20] The emphasis upon textual traditions that characterized so much early work in the study of religion and science has been particularly detrimental to the understanding of religion, science, and technology in nonwestern domains such as India. Attention to the experience of lived reality helps scholars overcome the limitations imposed by colonial history. This move toward postcolonial and nonwestern analysis—which is nearly a paradigm shift in itself—will productively redound upon the study of religion, science, and technology in western contexts also.

Furthermore, when scholarship concerns the lived reality of people, it gains relevance for decision-making by those people. For academic research to matter, it must be composed on the ground with the people who constitute its subject. The spectacular successes of digital technology in the late 20th and early 21st century and the potential for dramatic

---

[20] Drawing on the work of Gregory Schopen, I have elsewhere noted that this focus on texts and theories is a clear outgrowth of a Protestant emphasis upon 'the Word' which came to dominate academic thinking also (see Geraci, *Temples of Modernity*, 173–8).

improvements in AI behoove us to consider the political and economic implications of academic labors. This is not to disparage work in, for example, historical relationships between religion, science, and technology (much of which is fascinating in its own right and also bears upon the study of the present). Instead, it is a clarion call summoning scholars to new methodological practices and political emphases.

Greater intersection with the daily lives of stakeholder groups such as scientists, entrepreneurs, and the general public puts the study of religion, science, and technology in a direct relationship to wider culture. Its conclusions emerge out of engagement with that culture and are thus relevant within it. Scholars thus have a chance to be public-minded and a chance to reach out and show the importance of their work. Moving the study of religion, science, and technology beyond its early focus on conflict and harmony means we can pay attention to what the intersections of those domains actually does in society, which in turn means more substantive opportunities to contribute.[21]

Global visions of AI

This book explores the circulation of ideas about AI, not of the technologies themselves. It takes for granted that, barring disaster, technological development will continue, though it offers no projections on the directions that progress will take or the limits that constrain our technological outcomes. Without defining the ultimate potential of AI, this book nevertheless insists that our visions of AI powerfully influence the direction of progress. Apocalyptic perspectives emerged in the West through the historical confluence of transhumanist ideology, science fiction, and explosive growth in computation.[22] This apocalypticism thoroughly penetrates both insider and outsider views of AI and circulates globally. But the

---

[21] Elsewhere, I note that studying the practice of religious ritual in a scientific community 'indicates that rather than looking toward moments of intersection as proving grounds for theories of harmony, conflict, or anything in between, we learn more about both religion and science by thinking through what their points of intersection do, produce, or inform' ('Religious Ritual in a Scientific Space', 984).

[22] As noted earlier in this book, a number of political factors, most especially the broader eschatological views created by Cold War pressure and the potential of Earth-scale catastrophe through nuclear power, should be considered directly relevant as well.

global reception of this vision need not be uncritical welcome. Instead, we might look for the positive values it brings to discussions of AI and seek to leverage those while simultaneously recognizing that other cultural perspectives bring other worthwhile values. Indian values of duty and self-rule, for example, could be powerfully tied to the transformative and empowering elements of Apocalyptic AI to form a more globally valuable vision for the future.

Ideas about what AI can be are directly relevant to questions of AI design. This book notes some, though not all, contributions already extant from the U.S., and it begins an important conversation of how Indian values can be used as global values. But such a project does not stand in isolation. Humanity deserves an opportunity to think widely about the values that AI requires, both as a discipline of scientific and technological discovery and as a possible locus for a new species of intelligence. Seàn ÓeÉigertaigh and his colleagues note that this means a focus on cross-cultural cooperation.[23] This requires understanding and integrating philosophical and cultural resources from nonwestern cultures.

For example, Berberich, Nishida, and Suzuki provide a powerful contribution in their suggestion that Confucian understandings of harmony could lead to tactful AI that understands perspective and proportion in social activity. Their insightful play on words reveals how they aspire to 'make intelligent systems more harmonious ... but, at the same time, we have to be conscious of the mediating role that AI plays in our sociotechnical society. We should aim to build AI systems that help us to achieve better harmony between ourselves'.[24] Finding mechanisms for such social harmony will no doubt require that we find effective ways to express those values but also to exchange ideas on equal footing with one another. AI design, then, will be advanced by intercultural understanding that looks to leverage the achievements of multiple cultures, as exemplified in this book by the integration of ideas from India and the Euro-American West.

In an essay on India's early 20th-century cultural renaissance, Aurobindo Ghose (who would retire from politics and become the beloved spiritual leader Sri Aurobindo) provides instruction regarding

---

[23] ÓeÉigertaigh, et al., 'Overcoming Barriers to Cross-cultural Cooperation in AI Ethics and Governance'.

[24] Berberich, Nishida, and Suzuki, 'Harmonizing Artificial Intelligence for Social Good', 615.

the closure of gaps between India and the West. Aurobindo argues for a global and pluralistic vision of our human future, referring to 'the collective advance of the human race' and 'a collective advance towards the light, power, peace, unity, harmony of the diviner nature of humanity which the race is trying to evolve'.[25] There is no doubt that Aurobindo's sense of our evolutionary future differs in its spiritual content from the transhumanist perspective, but he and the iron horsemen share an expectation that transformation is in the works and that humanity can become something better. We saw a similar overlap in the published lectures of Haldane and Radhakrishnan. This overlap perhaps rescues AI from the individualism common in western transhumanism, producing instead a vision of humanity together.

To accomplish progress, Aurobindo argues that India must join Europe at the vanguard of progress rather than being forced to follow in Europe's wake (or, worse, choosing such a fate itself). 'It will be singular if while Europe is thus intelligently enlarging herself in the new light she has been able to seize and admitting the truths of the spirit and the aim at a divine change in man and his life [which Aurobindo credits to Europeans learning from India], we in India are to take up the cast-off clothes of European thought and life and to straggle along in the old rut of her wheels, always taking up today what she had cast off yesterday'.[26] For Aurobindo, India's renaissance required that Indians not merely follow along Europe some decades behind in thought, science, and culture. Instead, India must move in tandem with Europe, learning from what is new there and simultaneously contributing to the future. The same could and should be said of nations throughout the world.

We must attend to the frameworks occupied by technology because access and, more importantly, control is regulated by those contexts. Poonam Pandey and Aviram Sharma rightly acknowledge that in Indian conversations around technology access, that is, 'who will govern the resources, and how they will be governed, has been left out from the discussion'.[27] Obviously, this statement too can be broadened beyond India. As we assess the place of AI in global cultural contexts, we must be attentive to

---

[25] Ghosh, 'The Renaissance in India', 60, 63.

[26] Ghosh, 'The Renaissance in India', 63.

[27] Pandey and Sharma, 'Swinging between Responsibility and Rationality', 166.

the social locations of those technologies. Indeed, as Pandey and Sharma are at pains to demonstrate, public input matters when we judge the merits of a technology and its best deployment. In the case of AI, such a public is composed of many local and national communities, groups that should not be ignored in our race toward supposedly predetermined technical outcomes.

The struggles of India to establish itself politically, economically, and scientifically offer a valuable lesson of intercultural development. Our scientific, economic, and technological networks are global. Often, we fail to note how cultural ideas are likewise global, or that very possibility is suppressed in the political interests of certain groups. However, it is possible for us to think broadly about how parity in technological and cultural power can benefit humanity. Pankaj Sekhsaria notes of innovation and technology that:

> Different worldviews, knowledge systems and cultures of innovation are already at play, even if in a highly unequal world where many are forced to remain on the peripheries and have to struggle to survive. That many are not visible does not mean that they don't exist. Their invisibility is a function of the limited narratives and imagination of the so-called mainstream of science, technology, innovation, development and progress. Alternative narratives are possible—they already exist, in fact. All that is needed is for us to be sensitive to their possibilities and to be open to what they might have to offer.[28]

The openness of narrative brings with it an opportunity to greater reflection and superior resource management. Specifically, our ability to advance human life increases alongside our awareness and employment of cultural tools beyond those immediately available to any one culture. We must learn to draw on the intellectual, social, and ethical resources of many nations—both in understanding how they overlap and how they differ.

As we saw in previous chapters, 20th-century Indian thinkers shared the goal of cultural progress with Europeans; they also saw such progress as implicated in the growth of the human species. Following Rabindranath Tagore's evolutionary position that civilization implies

---

[28] Sekhsaria, *Instrumental Lives*, 78.

the end of nationhood and tribalism, K.C. Sen argues that people 'must be loved in groups' and that nationalism 'breeds war'.[29] Simultaneously, Haldane was arguing that 'it took man 250,000 years to transcend the hunting pack' but 'it will not take him so long to transcend the nation'.[30] Decades later, U.S.-based transhumanist Fereidoun Esfandiary, by then calling himself FM-2030, vigorously rejected national borders and considered them incompatible with transhumanist futures.[31] Commenting on Tagore, Sen notes that humankind shows little incentive toward a change of heart but that 'the intellect is subject to evolution. In its evolution lies the future of man' and suggests that eugenics—a spectre that haunts transhumanism[32]—could be necessary for human progress.[33]

In essence, Sen reads Tagore as being in touch with many of the very problematics that we would later assign to transhumanism. But Tagore's religious influences, from the Brahmo Samaj to the Baul Fakirs of Bengal, lead him away from scientific solutions. Nevertheless, Sen sums up, 'the ultimate aim of the god-man is to replace the race of men by a race of god-men'.[34] But of course neither Sen nor Tagore is anything that one might label transhumanist or even proto-transhumanist. Overlapping with Haldane and Huxley in the 20th century, they showed little interest in the technological transcendence favored by those two. Tagore and Sen show commitment to a future thoroughly integrated with traditional beliefs and the evolution to which they aspire is not one of technological transcendence but of moral transcendence. Political justice might be an outcome of scientific progress for Haldane, but it comes only through spiritual reform for Tagore and his allies.

Anachronistically, we might see a riposte to Kurzweil and the advocates of exponential progress in Tagore's thinking. Decades before the intelligence explosion of the Singularity was articulated—but simultaneous to the early modernists' adoration of social and technical

---

[29] Sen, 'The Religion of Man', 248. See also Tagore, 'Pathway to *Mukti*', 161–2.

[30] Haldane, *Daedalus*, 85.

[31] FM-2030, *Are You Transhuman?*, 132–4.

[32] President's Council on Bioethics, *Beyond Therapy*, 30–2; Tirosh-Samuelson, 'Engaging Transhumanism', 21–2, 24, 45–6. Perhaps unsurprisingly, Warwick even suggests that his proposed cyborg future would be one infused with eugenic principles: in his revelatory vision of the year 2050, 'becoming a cyborg is still through human birth and hence human babies are valued—the best of these being allowed to become cyborgs' (*I, Cyborg*, 303).

[33] Sen, 'The Religion of Man', 250, 252–3.

[34] Sen, 'The Religion of Man', 258.

acceleration—Tagore delivered his inaugural address as the first president of the Indian Philosophical Congress in which he stated:

> It should be realized that a mere addition to the rate of speed, to the paraphernalia of fast living and display of furniture, to the frightfulness of destructive armaments, only leads to an insensate orgy of a caricature of bigness. The links of bondage go on multiplying themselves, threatening to shackle the whole world with the chain forged by such unmeaning and unending urgency of need.[35]

For Tagore, humanity must work together, rejecting the desperate clinging of materialism and in doing so continuing the march of human progress. This is a progress of the soul—he writes that 'the true prayer of man is for the Real, not for the big, for the light which is not in incendiarism but in illumination, for Immortality which is not in duration of times but in the eternity of the perfect'[36]—and it speaks to his commitment to human flourishing. While this view is at first glance quite opposed to the Apocalyptic AI perspective, perhaps in that very opposition we can find a collaboration that brings out the best of both.

The translation and transformation of Apocalyptic AI and western technological imaginaries reflects a similar need to see the future as shared. Our contemplation of AI and our hope for human and environmental flourishing requires that multiple cultures bear the responsibility and share the outcomes of a global view of AI. Kimberly Hutchings notes that global ethics must be decolonized if it is to overcome the hegemonic control of any one culture.[37] If public consumption of AI and policy decisions about AI account only for apocalyptic visionaries wearing blinders that prevent other cultural contributions, then countries like India will be left following in a wake of destruction.

No single culture has all the answers, especially when it comes to technologies with the power of AI. Even if we never upload our minds into machines, we will wrestle with AI-empowered education, surveillance, military action, elder care, consumer capitalism, and professional

---

[35] Tagore, 'Pathway to *Mukti*', 160–1.
[36] Ibid., 164.
[37] Hutchings, 'Decolonizing Global Ethics'.

decision-making in medicine, law, and other domains. The values that urge us forward often go unnoticed and unremarked upon; it is time that the cultural and religious perspectives central to our technological imaginaries be fully uncovered and debated to produce a shared vision for AI. Cosmic transformations and optimistic dreams, mutual obligations of duty, promotion of self-sufficiency, and personal control are all potential tools to be leveraged. If we take such an obligation seriously, we may find that AI technologies become mighty participants in improving life for the marginalized, combatting climate change, and establishing just social structures.

References

Acquisti, Alessandro, Laura Brandimarte, and George Loewenstein. 2015. 'Privacy and Human Behavior in the Age of Information'. *Science* 347(6221): 509–514.

Acquisti, Alessandro, Ralph Gross, and Fred Stutzman. 2014. 'Face Recognition and Privacy in the Age of Augmented Reality'. *Journal of Privacy and Confidentiality* 6(2): 1–19.

Acquisti, Alessandro, Curtis Taylor, and Liad Wagman. 2016. 'The Economics of Privacy'. *Journal of Economic Literature* 54(2): 442–492.

Adas, Michael. 1989. *Machines as the Measure of Men: Science, Technology, and Ideologies of Western Dominance*. Ithaca: Cornell University Press.

Alba, Davey. 2019. 'The US Government Will Be Scanning Your Face at 20 Top Airports, Documents Show'. *Buzzfeed News*, March 11. Online. Available: https://www.buzzfeednews.com/article/daveyalba/these-documents-reveal-the-governments-detailed-plan-for (accessed August 18, 2019).

Albanese, Cathy. 1999. *America: Religions and Religion*. Belmont, CA: Wadsworth.

Alexander, Brian. 2003. *Rapture: How Biotech Became the New Religion*. New York: Basic Books.

Ali, Syed Mustafa. 2019. '"White Crisis" and/as "Existential Risk," or The Entangled Apocalypticism of Artificial Intelligence'. *Zygon: Journal of Religion and Science* 54(1): 207–224.

Ali, Syed Mustafa. 2021. 'Review of *Temples of Modernity* by Robert M. Geraci'. *Reading Religion*. Online. Available: https://readingreligion.org/books/temples-modernity

Altman, Michael J. 2017. *Heathen, Hindoo, Hindu: American Representations of India, 1721–1893*. New York: Oxford University Press.

Amarasingam, Amarnath. 2008. 'Transcending Technology: Looking at Futurology as a New Religious Movement'. *Journal of Contemporary Religion* 23(1): 1–16.

American Museum of Natural History. 2016. '2016 Isaac Asimov Memorial Debate: Is the Universe a Simulation?' Online. Available: https://www.youtube.com/watch?v=wgSZA3NPpBs (accessed May 27, 2019).

Angwin, Julia, Jeff Larson, Surya Mattu, and Lauren Kirchner. 2016. 'Machine Bias'. *ProPublica*, May 23. Online. Available: https://www.propublica.org/article/machine-bias-risk-assessments-in-criminal-sentencing (accessed July 2, 2020).

Apel, William D. 1979. 'The Lost World of Billy Graham'. *Review of Religious Research* 20(2): 138–149.

Appiah, Kwame Anthony. 2016. 'There Is No Such Thing as Western Civilization'. *The Guardian*, November 9. Online. Available: https://www.theguardian.com/world/2016/nov/09/western-civilisation-appiah-reith-lecture (accessed April 23, 2019).

Arnold, David. 2013. *Everyday Technology: Machines and the Making of India's Modernity*. Chicago: University of Chicago Press.

Asad, Talal. 1993. *Genealogies of Religion: Discipline and Reasons of Power in Christianity and Islam*. Baltimore: John Hopkins University Press.

Ashokananda. 1931. 'Science and the Future and the Future of Science'. *Prabuddha Bharat* 36(7): 316–323.

Asimov, Isaac. [1950] 1977. *I, Robot*. New York: Del Rey.

Asimov, Isaac. [1953] 1991. *The Caves of Steel*. New York: Bantam.

Asimov, Isaac. [1956] 1957. *The Naked Sun*. Garden City, NY: Doubleday.

Asimov, Isaac. [1983] 1991. *The Robots of Dawn*. New York: Del Rey.

Asprem, Egil. 2020. 'The Magus of Silicon Valley: Immortality, Apocalypse, and God Making in Ray Kurzweil's Transhumanism'. In *Mediality on Trial: Contesting and Contesting Trance and Other Media Techniques*, edited by Ehler Voss, pp. 397–411. Berlin: Walter de Gruyer GmbH.

Au, Wagner James. 2008. *The Making of Second Life: Notes From the New World*. New York: HarperCollins.

Au, Wagner James. 2011. 'In *The Terminator*, It's Skynet's Birthday—But Will Philip Rosedale Successfully Build a Real Skynet in Second Life?' *New World Notes*, April 19. Online. Available: https://nwn.blogs.com/nwn/2011/04/skynet-being-built-in-second-life.html (accessed April 5, 2019).

Au, Wagner James. 2019. 'Listen: Philip Rosedale Interviewed by Linden Vet about Building Virtual Worlds'. *New World Notes*, April 22. Online. Available: https://nwn.blogs.com/nwn/2019/04/philip-rosedale-linden-lab-high-fidelity-social-vr.html (accessed April 23, 2019).

Baber, Zaheer. 1996. *The Science of Empire: Scientific Knowledge, Civilization, and Colonial Rule in India*. Albany, NY: State University of New York Press.

Bacon, Francis. [1627] 1951. 'New Atlantis'. In *The Advancement of Learning and New Atlantis*, pp. 235–275. London: Oxford.

Bainbridge, William Sims. 2020. *Cultural Science: Applications of Artificial Social Intelligence*. New York: Business Expert Press.

Bajpai, G.S. and Mohsina Irshad. 2019. 'Artificial Intelligence, the Law and the Future'. *The Hindu*, June 11. Online. Available: https://www.thehindu.com/opinion/op-ed/artificial-intelligence-the-law-and-the-future/article27766446.ece (accessed August 18, 2019).

Balagangadhara, S.N. 2012. *Reconceptualizing India Studies*. New Delhi: Oxford University Press.

Balaram, P. 2007. 'The Promise of Biology and Biotechnology'. In *Science in India: Past and Present*, edited by B.V. Subbarayappa, pp. 409–427. Mumbai: Nehru Center and Popular Prakashan.

Baliga, Vrinda. 2018. 'The Collision of Parallels'. In *The Best Asian Speculative Fiction*, edited by Rajat Chaudhuri, pp. 322–332. Singapore: Kitaab.

Banerjee-Dube, Ishita. 2007. 'Reading Time'. In *Historical Anthropology*, edited by Saurabh Dube, pp. 149–164. New Delhi: Oxford University Press.

Banerjee-Dube, Ishita. 2015. *A History of Modern India*. Delhi: University of Cambridge Press.

Barbour, Ian. 1997. *Religion and Science: Historical and Contemporary Issues*. San Francisco: Harper-Collins.

Barker, Sean. 1987. 'Edmund Furse's The Theology of Robots'. *New Blackfriars* 68(801): 41–43.

Bartlett, Robert, Adair Morse, Richard Stanton, and Nancy Wallace. 2019. 'Consumer-Lending Discrimination in the FinTech Era'. National Bureau of Economic Research

Working Paper 25943. Online. Available: http://faculty.haas.berkeley.edu/morse/research/papers/discrim.pdf (accessed July 2, 2020).

Barrow, John D. and Frank Tipler. 1988. *The Anthropic Cosmological Principle*. New York: Oxford University Press.

Basalla, George. 1967. 'The Spread of Western Science'. *Science* 156(3775): 611–622.

Basu, Biman. 2009. 'Recent Developments in Science and Technology'. *Dream 2047* 11(12): 31–29 (reverse pagination).

Basu, Samit. 2010. *Turbulence*. Gurgaon: Hachette India.

Basu, Samit. 2014. *Resistance*. London: Titan.

Basu, Subho. 1998. 'Strikes and "Communal" Riots in Calcutta in the 1890s: Industrial Workers, Bhadralok Nationalist Leadership and the Colonial State'. *Modern Asian Studies* 32(4): 949–983.

BBC News. 2019. 'Elon Musk and Jack Ma Disagree about AI's Threat'. *BBC News*, August 29. Online. Available: https://www.bbc.com/news/technology-49508091 (accessed October 25, 2019).

Becque, Elien Blue. 2013. 'Elon Musk Wants to Die on Mars'. *Vanity Fair*, March 10. Online. Available: https://www.vanityfair.com/news/tech/2013/03/elon-musk-die-mars (accessed August 29, 2020).

Behe, Michael J. 1998. 'Molecular Machines: Experimental Support for the Design Inference'. *Cosmic Pursuit* 1(2): 27–35.

Behera, S.N. 2006. 'Science and Technology of Nanomaterials'. *Dream 2047* 8(8): 34–27 (reverse pagination).

Bendix, Aria. 2019. 'An Oxford Philosopher Who's Inspired Elon Musk Thinks Mass Surveillance Might Be the Only Way to Save Humanity from Doom'. *Business Insider*, April 19. Online. Available: https://www.businessinsider.com/nick-bostrom-mass-surveillance-could-save-humanity-2019-4 (accessed August 18, 2019).

Benedikt, Michael (ed.). 1991. *Cyberspace; First Steps*. Cambridge, MA: The MIT Press.

Berberich, Nicolas, Toyoaki Nishida, and Shoko Suzuki. 2020. 'Harmonizing Artificial Intelligence for Social Good'. *Philosophy & Technology* 33(4): 613–638.

Berman, John. 2011. 'Futurist Ray Kurzweil Says He Can Bring His Dead Father Back to Life Through a Computer Avatar'. ABC News, August 9. Online. Available: https://abcnews.go.com/Technology/futurist-ray-kurzweil-bring-dead-father-back-life/story?id=14267712 (accessed April 2, 2019).

Bernstein, Anya. 2015. 'Freeze, Die, Come to Life: The Many Paths to Immortality in Post-Soviet Russia'. *American Ethnologist* 42(4): 766–81.

Bernstein, Anya. 2019. *The Future of Immortality: Remaking Life and Death in Contemporary Russia*. Princeton, NJ: Princeton University Press.

Bernstein, Anya. 2019. 'Life Unlimited: Russian Archives of the Digital and the Human'. *Journal of the Royal Anthropological Institute* 25(4): 676–697.

Bardhan, Adrish. 2019. 'Planet of Terror'. In *The Gollancz Book of South Asian Science Fiction*, edited by Tarun K. Saint and Manjula Padmanabhan, pp. 1–6. Gurugram: Hachette India.

Bhargava, Rajeev. 2016. 'An Ancient Indian Secular Age?' In *Beyond the Secular West*, edited by Akeel Bilgrami, pp. 188–214. New York: Columbia University Press.

Bhatia, C.R. 2018. 'CRISPR—Cas9: The Latest Tool for Genetic Enhancement of Crop Plants'. *Dream 2047* 21(3): 23–22 (reverse pagination).

Bhattacharjee, Govind. 2018. 'Age of Man-Machine Hybrids'. *Dream 2047* 21(2): 30–26 (pagination runs in reverse due to dual-language publication).

Bhattacharjee, Govind. 2018. 'The Curious Case of Unruly Robots'. *Science Reporter* 55(10): 14–19.

Bhattacharyya, K.C. [1928] 2011. 'Savaraj in Ideas'. In *Indian Philosophy in English: From Renaissance to Independence*, edited by Nalini Bhushan and Jay L. Garfield, pp. 101–114. New York: Oxford University Press.

Bhushan, Nalini and Jay L. Garfield. 2017. *Minds Without Fear: Philosophy in the Indian Renaissance*. New York: Oxford University Press.

Bilefsky, Dan. 2019. 'He Helped Create A.I. Now, He Worries about "Killer Robots"'. *The New York Times*, March 29. Online. Available: https://www.nytimes.com/2019/03/29/world/canada/bengio-artificial-intelligence-ai-turing.html (accessed August 18, 2019).

Bilmoria, Purushottama Bilimoria. 2007. 'A Prolegomenon for All Future Dialogues Between Science and Religion or Spirituality in India'. In *Traditions of Science: Cross-Cultural Perspectives: Essays in Honour of B.V. Subbarayappa*, edited by Purshottama Bilimoria and Melukote K. Sridhar, pp. 210–221. New Delhi: Munshiram Manoharlal.

Binns, Reuben. 2018. 'Algorithmic Accountability and Public Reason'. *Philosophy & Technology* 31(4): 543–556.

Blascovich, Jim and Jeremy Bailenson. 2011. *Infinite Reality: Avatars, Eternal Life, New Worlds, and the Dawn of the Virtual Revolution*. New York: William Morrow.

Borges, Jorge Luis. [1962] 2007. 'Tlön, Uqbar, Orbis Tertius'. In *Labyrinths: Selected Stories and Other Writings*, edited by Donald A. Yates and James E. Irby, pp. 3–18. New York: New Directions.

Bose, Debarpita, Debasmita Das, Ishana Ghosh, Srishti Barua, and Nasima A. Mazarbhuiyan. 2020. 'Indianness?—A Complex Hybridization of Cultures in Samit Basu's "The Simoqin Prophecies"'. *Research Journal of English Language and Literature* 8(3): 332–344.

Bostrom, Nick. 2003. 'Are We Living in a Computer Simulation?' *The Philosophical Quarterly* 53(211): 243–255.

Bostrom, Nick. 2005. 'In Defense of Posthuman Dignity'. *Bioethics* 19(3): 202–214.

Bostrom, Nick. 2015. 'It's Still Early Days'. In *What to Think about Machines That Think*, edited by John Brockman, pp. 126–127. New York: Harper Perennial.

Boyer, Paul. 1992. *When Time Shall Be No More: Prophecy Belief in Modern American Culture*. Cambridge, MA: Harvard University Press.

Boyer, Paul. 2000. 'The Growth of Fundamentalist Apocalyptic in the United States'. In *The Encyclopedia of Apocalypticism, Volume 3, Apocalypticism in the Modern Period and the Contemporary Age*, edited by Stephen J. Stein, pp. 140–178. New York: Continuum.

Brockman, John (ed.). 2015. *What To Think About Machines That Think*. New York: Harper Perennial.

Brockman, John (ed.). 2019. 'Introduction: On the Promise and Peril of AI'. In *Possible Minds: Twenty-Five Ways of Looking at AI*, edited by John Brockman, pp. xv–xxvi. New York: Penguin.

Brooke, John Hedley and Ronald L. Numbers (ed.). 2011. *Science and Religion Around the World*. New York: Oxford University Press.

Brown, C. Mackenzie. 2012. *Hindu Perspectives on Evolution: Darwin, Dharma, and Design*. New York: Routledge.

Buben, Adam. 2019. 'Personal Immortality in Transhumanism and Ancient Indian Philosophy'. *Philosophy East & West* 69(1): 71–85.

Bull, Malcolm. 1999. *Seeing Things Hidden: Apocalypse, Vision and Totality*. New York: Verso.

Burdett, Michael S. 2011. 'Contextualizing a Christian Perspective on Transcendence and Human Enhancement: Francis Bacon, N.F. Fedorov, and Pierre Teilhard de Chardin'. In *Transhumanism and Transcendence: Christian Hope in an Age of Technological Enhancement*, edited by Ronald Cole-Turner, pp. 19–36. Washington, D.C.: Georgetown University Press.

Burdett, Michael S. 2015. *Eschatology and the Technological Future*. New York: Routledge.

Bush, Vannevar. 1945. 'As We May Think'. *The Atlantic*, July. Online. Available: https://www.theatlantic.com/magazine/archive/1945/07/as-we-may-think/303881/ (accessed October 24, 2018).

Butler, Philip. 2020. *Black Transhuman Liberation Theology: Technology and Spirituality*. New York: Bloomsbury.

Byerley, Melinda. 2019. 'Philip Rosedale: "Starman"'. *Stayin' Alive in Technology Podcast*, April 18. Online. Available: https://www.stayinaliveintech.com/podcast/2019/s2-e9/philip-rosedale-starman (accessed June 2, 2019).

Calo, Ryan. 2016. 'Can Americans Resist Surveillance?' *The University of Chicago Law Review* 83(1): 23–43.

Cave, Stephen and Kanta Dihal. 2019. 'Hopes and Fears for Intelligent Machines in Fiction and Reality'. *Nature Machine Intelligence* 1(2): 74–78.

Cave, Stephen and Kanta Dihal. 2020. 'The Whiteness of AI'. *Philosophy & Technology* 33(4): 685–703.

Cave, Stephen, Kanta Dihal, and Sarah Dillon. 2020. 'Introduction: Imagining AI'. In *AI Narratives: A History of Imaginative Thinking About Intelligent Machines*, edited by Cave, Dihal, and Dillon, pp. 1–21. New York: Oxford University Press.

Cellan-Jones, Rory. 2014. 'Stephen Hawking Warns Artificial Intelligence Could End Mankind'. *BBC.com*, December 2. Online. Available: https://www.bbc.com/news/technology-30290540 (accessed April 2, 2019).

Chabria, Priya Sarukkai. 2008. *Generation 14*. New Delhi: Zubaan.

Chakbrabarti, Pratik. [2004] 2010. *Western Science in Modern India: Metropolitan Methods, Colonial Practices*. Ranikhet: Permanent Black.

Chakbrabarti, Pratik. 2012. *Bacteriology in British India: Laboratory Medicine and the Tropics*. Rochester: University of Rochester Press.

Chandra, Gita. 2019. 'The Goddess Project'. In *The Gollancz Book of South Asian Science Fiction*, edited by Tarun K. Saint and Manjula Padmanabhan, pp. 221–235. Gurugram: Hachette India.

Chatterjee, Partha. [1986] 1999. 'Nationalist Thought and the Colonial World'. In *The Partha Chatterjee Ombnibus*. New Delhi: Oxford University Press.

Chatterjee, Partha. 1992. 'History and the Nationalization of Hinduism'. *Social Research* 59(1): 111–149.

Chatterjee, Partha. [1993] 1999. *The Nation and Its Fragments: Colonial and Postcolonial Histories*. In *The Partha Chatterjee Omnibus*. New Delhi: Oxford University Press.

Chattopadyay, Kamaladevi. [1980] 2000. *India's Craft Tradition*. New Delhi: Government of India Ministry of Information and Broadcasting.

Chaudhary, Mohammad Yaqub. 2019. 'Augmented Reality, Artificial Intelligence, and the Re-Enchantment of the World'. *Zygon: Journal of Religion and Science* 54(2): 454–478.

Chaudhry, Lakshmi. 2000. 'Valley to Bill Joy: "Zzzzzzz"'. *Wired*, April 5. Online. Available: https://www.wired.com/2000/04/valley-to-bill-joy-zzzzzzz/ (accessed June 4, 2019).

Chawla, Neeraj. 2018. 'The God Link'. In *The Best Asian Speculative Fiction*, edited by Rajat Chaudhuri, pp. 343–359. Singapore: Kitaab.

Cheong, Pauline Hope. 2020. 'Religion, Robots and Rectitude: Communicative Affordances for Spiritual Knowledge and Community'. *Applied Artificial Intelligence* 34(5): 412–431.

Chidester, David. 1996. *Savage Systems: Colonialism and Comparative Religion in Southern Africa*. Charlottesville: University of Virginia Press.

Chidester, David. 2014. *Empire of Religion: Imperialism and Comparative Religion*. Chicago: University of Chicago Press.

Clarke, Arthur C. 1953. *Childhood's End*. New York: Balantine.

Clarke, Arthur C. 1968. *2001: A Space Odyssey*. New York: New American Library.

Clarke, Arthur C. [1956] 2001. *The City and the Stars*. In *The City and the Stars and The Sands of Mars*. New York: Warner.

Clayton, Jay. 2013. 'The Ridicule of Time: Science Fiction, Bioethics, and the Posthuman'. *American Literary History* 25(2): 317–343.

Cole-Turner, Ronald. 2015. 'Going Beyond the Human: Christians and Other Transhumanists'. *Theology and Science* 13(2): 150–161.

Colors Magazine. 1994. 'The Buddhist Monk Machine'. *Colors: A Magazine about the Rest of the World* 8: 33.

Collins, John J. 1984. *The Apocalyptic Imagination: An Introduction to the Jewish Matrix of Christianity*. New York: Crossroad.

Collins, John J. 2000. 'From Prophecy to Apocalypticism: The Expectation of the End'. In *The Encyclopedia of Apocalypticism*, edited by Bernard McGinn, John J. Collins, and Stephen J. Stein, Vol. 1, *The Origins of Apocalypticism in Judaism and Christianity*, edited by John J. Collins, pp. 129–161. New York: Continuum.

Cookson, Clive and Patrick McGee. 2019. 'Cyborgs: Elon Musk and the New Era of Neuroscience'. *Financial Times*, July 19. Online. Available: https://www.ft.com/content/eec3bfb2-aa09-11e9-b6ee-3cdf3174eb89 (accessed August 30, 2020).

Cramer, Benjamin W. 2018. 'A Proposal to Adopt Data Discrimination Rather than Privacy as the Justification for Rolling Back Data Surveillance'. *Journal of Information Policy* 8: 5–33.

Crevier, Daniel. 1993. *AI: The Tumultuous History of the Search for Artificial Intelligence*. New York: Basic Books.

D'Agostino, Marcello and Massimo Durante. 2018. 'Introduction: The Governance of Algorithms'. *Philosophy & Technology* 31(4): 499–505.

Dalal, Pankti. 2014. 'Gujarat Model of Using Epics as Facts in Education'. DNA India, July 27. Online. Available: https://www.dnaindia.com/india/report-gujarat-model-of-using-epics-as-facts-in-education-2005721 (accessed July 22, 2019).

Danaher, John. 2018. 'Towards an Ethics of AI Assistants: An Initial Framework'. *Philosophy & Technology* 31(4): 629–653.

Daniel, E. Valentine. 2000. 'The Arrogation of Being: Revisiting the Anthropology of Religion'. *Macalester International* 8 (1): 171–191.

Das, Indrapramit. 2012. 'Sita's Descent'. In *Breaking the Bow: Speculative Fiction Inspired by The Ramayana*, edited by Anil Menon and Vandana Singh, pp. 105–114. New Delhi: Zubaan.

Das, Jayashree, Pritha Dey, and Pradipta Banerjee. 2018. 'Crisp and CRISPR: Designing Our Descendants'. *Dream 2047* 20(8): 25–22 (reverse pagination).

Dawesar, Abha. 2012. 'The Good King'. In *Breaking the Bow: Speculative Fiction Inspired by The Ramayana*, edited by Anil Menon and Vandana Singh, pp. 45–61. New Delhi: Zubaan.

De Garis, Hugo. 2005. *The Artilect War: Cosmists vs. Terrans: A Bitter Controversy Concerning Whether Humanity Should Build Godlike Massively Intelligent Machines*. Palm Springs, CA: ETC Publications.

De Roover, Jakob. 2019. 'A New School in the Study of India?' *Contemporary South Asia* 27(2): 273–285.

Deese, R.S. 2015. *We Are Amphibians: Julian and Aldous Huxley on the Future of Our Species*. Oakland: University of California Press.

Delio, Ilia. 2020. 'Religion and Posthuman Life: A Note on Teilhard de Chardin's Vision'. *Toronto Journal of Theology*: Advance online access.

DeNapoli, Antoinette. 2017. '"Dharm" is Technology": The Theologizing of Technology in the Experimental Hinduism of Renouncers in Contemporary North India'. *International Journal of Dharma Studies* 5:18.

Dennett, Daniel. 2015. 'The Singularity—An Urban Legend?' In *What To Think about Machines That Think*, edited by John Brockman, pp. 85–92. New York: Harper Perennial.

Dennett, Daniel. 2019. 'What Can We Do?' In *Possible Minds: Twenty-Five Ways of Looking at AI*, edited by John Brockman, pp. 41–53. New York: Penguin.

Deshpande, Madhav. 1979. 'History, Change and Permanence: A Classical Indian Perspective'. In *Contributions to South Asian Studies 1*, edited by Gopal Krishna, pp. 1–28. Delhi: Oxford University Press.

Deutsch, David. 2019. 'Beyond Reward and Punishment'. In *Possible Minds: Twenty-Five Ways of Looking at AI*, edited by John Brockman, pp. 113–124. New York: Penguin.

Diamond, Jared. 1997. 'The Curse of QWERTY'. *Discover*, April 1. Online. Available:https://www.discovermagazine.com/technology/the-curse-of-qwerty(accessed October 28, 2018).

Doniger, Wendy. 2014. *On Hinduism*. New York: Oxford University Press.

Dorit, Robert L. 2009. 'Marginalia: Keyboards, Codes and the Search for Optimality'. *American Scientist* 97(5): 376–9.

Drexler, Eric K. [1986] 1990. *Engines of Creation*. New York: Anchor.

Dronamraju, Krishna. 1995. 'Introduction'. In *Haldane's Daedalus Revisited*, edited by Krishna Dronamraju, pp. 1–22. New York: Oxford University Press.

Dronamraju, Krishna (ed.). 2009. *What I Require from Life: Writings on Science and Life from J.B.S. Haldane*. Oxford: Oxford University Press.

Dronamraju, Krishna. 2010. 'J.B.S. Haldane's Last Years: His Life and Work in India (1957–1964)'. *Genetics* 185(1): 5–10.

Dronamraju, Krishna. 2019. Email correspondence with the author. June 1, 2019.

Ecklund, Elaine Howard, David R. Johnson, Brandon Vaidyanthan, Kristin R.W. Matthews, Steven W. Lewis, Robert A. Thomson, and Di Di. 2019. *Secularity and Science: What Scientists Around the World Really Think About Religion*. New York: Oxford.

Ecklund, Elaine Howard, David R. Johnson, Christopher P. Scheitle, Kirstin R.W. Matthews, and Steven W. Lewis. 2016. 'Religion Among Scientists in International Context: A New Study of Scientists in Eight Regions'. *Socius: Sociological Research for a Dynamic World* 2: 1–9.

Esfandiary, Fereidoun. 'The Mystical West Puzzles the Practical East'. *The New York Times*, February 5, p. SM112.

Esfandiary, Fereidoun. [1970] 1978. *Optimism One*. New York: Popular Library.

Esfandiary, Fereidoun. [1973] 1977. *Up-Wingers: A Futurist Manifesto*. New York: Popular Library.

Esfandiary, Fereidoun. 1981. 'Up-Wing Priorities'. *Future Life* 27(June): 70–73.

Etherington, Darrell. 2019. 'Elon Musk's Neuralink looks to Begin Outfitting Human Brains with Faster Input and Output Starting Next Year'. *TechCrunch*, July 16. Online. Available: https://techcrunch.com/2019/07/16/elon-musks-neuralink-looks-to-begin-outfitting-human-brains-with-faster-input-and-output-starting-next-year/ (accessed August 30, 2020).

Ethics & Religious Liberty Commission. 2019. 'Artificial Intelligence: An Evangelical Statement of Principles'. *Southern Baptist Convention*, April 11. Online. Available: https://erlc.com/resource-library/statements/artificial-intelligence-an-evangelical-statement-of-principles (accessed May 27, 2019).

Ettinger, Robert. 1964. *The Prospect of Immortality*. New York: Doubleday & Company.

Ettinger, Robert. 1972. *Man into Superman*. New York: Avon.

Evans, Dylan. 2015. 'The Great AI Swindle'. In *What to Think About Machines That Think*, edited by John Brockman, pp. 209–211. New York: Harper Perennial.

Faggella, Daniel. 2018. '(All) Elon Musk Artificial Intelligence Quotes – A Catalogue of His Statements'. *Emerj.com*, November 29. Online. Available: https://emerj.com/ai-future-outlook/elon-musk-on-the-dangers-of-ai-a-catalogue-of-his-statements/ (accessed April 4, 2019).

Farman, Abou. 2019. 'Mind out of Place: Transhuman Spirituality'. *Journal of the American Academy of Religion* 87(1): 57–80.

Feakins, Paul. 2017. 'The Singularity is Near: How Kurzweil's Predictions are Faring'. *Antropy: Ecommerce Experts website*, January 29. Online. Available: https://www.antropy.co.uk/blog/the-singularity-is-near-how-kurzweils-predictions-are-faring/ (accessed November 15, 2018).

Feenberg, Andrew. 1992. 'Subversive Rationalization: Technology, Power, and Democracy'. *Inquiry* 35(3/4): 301–322.

Fehige, Yiftach (ed.). 2016. *Science and Religion: East and West*. New York: Routledge.

Fisher, Matthew Zaro. 2015. 'More Human than the Human? Toward a 'Transhumanist' Christian Theological Anthropology'. In *Religion and Transhumanism: The Unknown Future of Human Enhancement*, edited by Calvin R. Mercer and Tracy J. Trothen, pp. 23–38. Santa Barbara, CA: ABC-CLIO.

Foerst, Anne. 1998. 'Cog, a Humanoid Robot, and the Question of the Image of God'. *Zygon: Journal of Religion and Science* 33 (1): 91–111.

Foerst, Anne. 2004. *God in the Machine: What Robots Teach Us About Humanity and God*. New York: Dutton.

Foucault, Michel. [1966] 1994. *The Order of Things: An Archeology of the Human Sciences*. New York: Vintage.

Foucault, Michel. [1969] 1972. *The Archaeology of Knowledge and The Discourse on Language*, translated by A.M. Sheridan Smith. New York: Pantheon.

Foucault, Michel. [1975] 1995. *Discipline and Punish: The Birth of the Prison*, translated by Alan Sheridan. New York: Vintage.

Freud, Sigmund. [1923] 1960. *The Ego and the Id*. New York: W.W. Norton.

Freud, Sigmund. [1913] 1990. *Totem and Taboo*. New York: W.W. Norton.

Furse, Edmund. 1986. 'The Theology of Robots'. *New Blackfriars* 67(795): 377–386.

Furse, Edmund. 1996. 'Towards the First Catholic Robot?' *The Independent*, October 25. Online. Available: https://www.independent.co.uk/voices/letters-towards-the-first-catholic-robot-1360242.html (accessed July 15, 2021).

Gallacher, John D., Monica Kaminska, Bence Kollanyi, and Philip N. Howard. 2017. 'Junk News and Bots during the 2017 UK General Election: What Are UK Voters Sharing over Twitter?' COMPROP Data Memo 2017.5 Working Paper. Online. Available: https://comprop.oii.ox.ac.uk/research/working-papers/junk-news-and-bots-during-the-2017-uk-general-election/ (accessed July 8, 2020).

Gandhi, Mohandas K. [1909] 1922. *Indian Home Rule*. Madras: Ganesh & Co. Kindle version.

Garner, Stephen. 2011. 'The Hopeful Cyborg'. In *Transhumanism and Transcendence: Christian Hope in an Age of Technological Enhancement*, edited by Ronald Cole-Turner, pp. 87–100. Washington, D.C.: Georgetown University Press.

Gayle, Damien. 2019. 'UK, US and Russia Among Those Opposing Killer Robot Ban'. *The Guardian*, March 29. Online. Available: https://www.theguardian.com/science/2019/mar/29/uk-us-russia-opposing-killer-robot-ban-un-ai (accessed March 29, 2019).

Geduld, Harry M. and Ronald Gottesman (eds.). 1978. *Robots, Robots, Robots*. Boston: New York Graphic Society.

Geraci, Robert M. 2006. 'Spiritual Robots: Religion and Our Scientific View of the Natural World'. *Theology and Science* 4(3): 229–246.

Geraci, Robert M. 2007. 'Cultural Prestige: Popular Science Robotics as Religion-Science Hybrid'. In *Reconfigurations: Interdisciplinary Perspectives on Religion in a Post-Secular Society*, edited by Alexander Ornella and Stefanie Knauss, pp. 43–58. Vienna: LIT Press.

Geraci, Robert M. 2007. 'Robots and the Sacred in Science and Science Fiction: Theological Reflections on Artificial Intelligence'. *Zygon: Journal of Religion and Science* 42(4): 961–980.

Geraci, Robert M. 2008. 'Apocalyptic AI: Religion and the Promise of Artificial Intelligence'. *Journal of the American Academy of Religion* 76(1): 138–166.

Geraci, Robert M. 2010. *Apocalyptic AI: Visions of Heaven in Robotics, Artificial Intelligence, and Virtual Reality*. New York: Oxford University Press.

Geraci, Robert M. 2010. 'The Popular Appeal of Apocalyptic AI'. *Zygon: Journal of Religion and Science* 45(4): 1003–1020.

Geraci, Robert M. 2011. 'There and Back Again: Transhumanist Evangelism in Science Fiction and Popular Science'. *Implicit Religion* 14(2): 141–172.

Geraci, Robert M. 2012. 'Video Gaming and the Transhuman Inclination'. *Zygon: Journal of Religion and Science* 47(4): 735–756.

Geraci, Robert M. 2016. 'L'Évangélisme Transhumaniste'. In *PERSONA Exhibition Catalog*, edited by E. Grimaud, pp. 212–213. Paris: Musée du Quai Branley.

Geraci, Robert M. 2018. 'Saffron Glasses: Indian Nationalism and the Enchantment of Technology'. In *Religion and Technology in India: Spaces, Practices and Authorities*, edited by Knut A. Jacobsen and Kristina Myrvold, pp. 25–42. New York: Routledge.

Geraci, Robert M. 2018. *Temples of Modernity: Nationalism, Hinduism, and Transhumanism in South Indian Science*. Lanham, MD: Lexington.

Geraci, Robert M. 2019. 'Religion, Technology, and Human-Computer Interaction'. Presented at the Srishti School of Art, Design & Technology, Association for Computing Machinery SIGCHI Student Chapter; Bangalore, India (March 27). Online. Available: https://www.youtube.com/watch?v=B74_sC556zU&t=5s (accessed October 25, 2019).

Geraci, Robert M. 2019. 'Religious Ritual in Scientific Spaces: Festival Participation and the Integration of Outsiders'. *Science, Technology & Human Values* 44(6): 965–993.

Geraci, Robert M. 2020. 'A Hydra-logical Approach: Acknowledging Complexity in the Study of Religion, Science, and Technology'. *Zygon: Journal of Religion and Science* 55(4): 948–970.

Geraci, Robert M and Yong Sup Song. Forthcoming. 'Global Culture for Global Technology: Religious Values and Progress in Artificial Intelligence'.

Gershenfeld, Neil. 2019. 'Scaling'. In *Possible Minds: Twenty-Five Ways of Looking at AI*, edited by John Brockman, pp. 160–169. New York: Penguin.

Ghosh, Aurobindo. [1918] 2011. 'The Renaissance in India'. In *Indian Philosophy in English: From Renaissance to Independence*, edited by Nalini Bhushan and Jay L. Garfield, pp. 37–65. New York: Oxford University Press.

Ghosh, Jogendra Chandra. 1937. 'Aśvameda and Rājasūya'. *Indian Culture* 3(4): 763–764.

Ghosh, Sujoy. 2017. *Anukul*. India: LargeShortFilms. Online. Available: https://www.youtube.com/watch?v=J2mqIgdae5I (accessed March 12, 2019).

Gibson, William. 1984. *Neuromancer*. New York: Ace.

Giddings, Seth. 2019. 'Toying with the Singularity: AI, Automata and Imagination in Play with Robots and Virtual Pets'. In *The Internet of Toys: Practices, Affordances and the Political Economy of Children's Smart Play*, edited by Giovanna Mascheroni and Donell Holloway, pp. 67–87. New York: Palgrave Macmillan.

Gill, Amandeep Singh. 2019. 'Artificial Intelligence and International Security: The Long View'. *Ethics & International Affairs* 33(2): 169–179.

Giri, Sri Yukteswar. [1894] 1972. *The Holy Science*. Los Angeles: Self-Realization Fellowship.

Giuliano, Roberto Musa. 2020. 'Echoes of Myth and Magic in the Language of Artificial Intelligence'. *AI & Society*: Advance online access.

Gladden, Matthew E. 2019. 'Who Will Be the Members of Society 5.0? Towards an Anthropology of Technologically Posthumanized Future Societies'. *Social Sciences* 8(5): 1–39.

Goertzel, Ben. 2010. *A Cosmit Manifesto*. Los Angeles: Humanity+ Press.

Gold, Ann. 1998. 'Sin and Rain, Moral Ecology in Rural North India'. In *Purifying the Earthly Body of God: Religion and Ecology in Hindu India*, edited by Lance E. Nelson, pp. 165–196. Albany: SUNY Press.

Goldman, Andrew. 2013. 'Ray Kurzweil Says We're Going to Live Forever'. *The New York Times*, January 25. Online. Available: http://www.nytimes.com/2013/01/27/magazine/ray-kurzweil-says-were-going-to-live-forever.html

González-Reimann, Louis. 2002. *The Mahābhārata and the Yugas: India's Great Epic Poem and the Hindu System of World Ages*. New York: Peter Lang.

González-Reimann, Louis. 2013. 'The Coming Golden Age: On Prophecy in Hinduism'. In *Prophecy in the New Millennium: When Prophecies Persist*, edited by Sarah Harvey and Suzanne Newcombe, pp. 105–122. Burlington: Ashgate.

Good, Irving. 1966. 'Speculations Concerning the First Ultra Intelligent Machine'. *Advances in Computers* 6: 31–88.

Gopal, Sarvepalli. 1989. *Radhakrishnan: A Biography*. Delhi: Oxford University Press. Archived reprint available at https://archive.org/stream/in.ernet.dli.2015.131084/2015.131084.Radhakrishnan-A-Biography_djvu.txt (accessed April 2, 2020).

Gopnik, Alison. 2019. 'AIs Versus Four-Year-Olds'. In *Possible Minds: Twenty-Five Ways of Looking at AI*, edited by John Brockman, pp. 219–230. New York: Penguin.

Gosling, David L. 2007. *Science and the Indian Tradition: When Einstein Met Tagore*. New York: Routledge.

Gottschalk, Peter. 2013. *Religion, Science, and Empire: Classifying Hinduism and Islam in British India*. New York: Oxford University Press.

Gould, Hannah and Holly Walters. 2020. 'Bad Buddhists, Good Robots: Techno-Salvationist Designs for Nirvana'. *Journal of Global Buddhism* 21: 277–294.

Gould, Stephen Jay. 1999. *Rocks of Ages: Science and Religion in the Fullness of Life*. New York: The Library of Contemporary Thought.

Government of Japan. 2016. *The 5th Science and Technology Basic Plan (Provisional Translation)*. Tokyo: Council for Science, Technology and Innovation. Online. Available: https://www8.cao.go.jp/cstp/english/basic/5thbasicplan.pdf (accessed May 13, 2019).

Grace, Katja, John Salvatier, Allan Dafoe, Baobao Zhang, and Owain Evans. 2018. 'Viewpoint: When Will AI Exceed Human Performance? Evidence from AI Experts'. *Journal of Artificial Intelligence Research* 62: 729–754.

Graham, Elaine. 2002. '"Nietzsche Gets a Modem": Transhumanism and the Technological Sublime'. *Literature and Theology* 16(1): 65–80.

Grassegger, Hannes and Mikael Krogerus. 2018. 'The Data that Turned the World Upside Down'. *Vice.com*, March 17 (updated from January 28, 2017). Online. Available: https://www.vice.com/en_us/article/mg9vvn/how-our-likes-helped-trump-win (accessed July 8, 2020).

Gray, Chris Hables. 1997. 'Artificial Intelligence at War: An Analysis of the Aegis System in Combat'. In *Reinvinenting Technology, Redisocvering Community: Critical

Explorations of Computing as a Social Practice, edited by Philip E. Agre and Douglas Schuler, pp. 127–142. Greenwich, CT: Ablex.

Green, Brian Patrick. 2018. 'The Technology of Holiness: A Response to Hava Tirosh-Samuelson'. *Theology and Science* 16(2): 223–228.

Green, Erin Elizabeth. 2018. *Robots and AI: The Challenge to Interdisciplinary Theology*. Doctoral dissertation submitted to Graduate Centre for Theological Studies of the Toronto School of Theology and the University of St. Michael's College, University of Toronto.

Greenberg, David S. [1967] 1999. *The Politics of Pure Science*. Chicago: University of Chicago Press.

Griffiths, Tom. 2019. 'The Artificial Use of Human Beings'. In *Possible Minds: Twenty-Five Ways of Looking at AI*, edited by John Brockman, pp. 125–133. New York: Penguin.

Grossman, Lev. 2011. '2045: The Year Man Becomes Immortal'. *Time*, February 10. Online. Available: http://content.time.com/time/magazine/article/0,9171,2048299,00.html (accessed September 23, 2017).

Groys, Boris (ed.). 2018. *Russian Cosmism*. Cambridge, MA: The M.I.T. Press.

Guest, Tim. 2007. *Second Lives: A Journey through Virtual Worlds*. New York: Random House.

Guha, Ranajit. 1999. *Elementary Aspects of Peasant Insurgency in Colonial India*. Durham, NC: Duke University Press.

Gupta, Dipankar. 2009. *The Caged Phoenix: Can India Fly?* New Delhi: Viking.

Gupta, Subhojoy. 2018. 'Icarus'. In *An Anthology of Indian Sci-Fi Stories*, produced by the Bangalore Sci-Fi Club. Singapore: Amazon Asia-Pacific Holdings. Kindle publication.

Gurukkal, Rajan. 2014. 'A Blindness About India'. *Economic & Political Weekly* 49(December 6): 12–15.

Habib, S. Irfan and Dhruv Raina. 1989. 'Copernicus, Colombus, Colonialism and the Role of Science in Nineteenth Century India'. *Social Scientists* 17(3/4): 51–66.

Haldane, J.B.S. 1924. *Daedalus, Or Science and the Future*. New York: E.P. Dutton.

Haldane, J.B.S. 1952. 'A Biologist Looks at India'. *The Visvabharati Quarterly* 17(4): 271–281.

Haldane, J.B.S. 1952. Untitled greetings. *Indian Rationlist* 1(1): 5.

Haldane, J.B.S. 1958. 'Message from J.B.S. Haldane'. *Indian Rationalist* 6(5/6): 2, 15.

Haldane, J.B.S. [1959] 2009. 'Simplifying Astronomy'. In *What I Require from Life: Writings on Science and Life from J.B.S. Haldane*, edited by Krishna Dronamraju, pp. 191–194. New York: Oxford University Press.

Haldane, J.B.S. [1959] 2009. 'Some Reflections on Non-Violence'. In *What I Require from Life: Writings on Science and Life from J.B.S. Haldane*, edited by Krishna Dronamraju, pp. 132–139. New York: Oxford University Press.

Haldane, J.B.S. 1963. 'Biological Possibilities In the Next Ten Thousand Years'. In *Man and His Future*, edited by Gordon Wolstenholme, pp. 337–361. Boston: Little, Brown and Company.

Hansen, Thomas Blom. 1999. *The Saffron Wave: Democracy and Hindu Nationalism in Modern India*. Princeton, NJ: Princeton University Press.

Harari, Noah Yuval. [2011] 2015. *Sapiens: A Brief History of Humankind*. New York: Harper.

Haraway, Donna. 1997. *Modest_Witness@Second_Millennium.FemaleMan©_Meets_ Oncomouse™: Feminism and Technoscience*. New York: Routledge.

Harding, Sandra. 1988. *Is Science Multicultural? Postcolonialisms, Feminisms, and Epistemologies*. Bloomington: Indiana University Press.

Harris, Sam. 2015. 'Can We Avoid a Digital Apocalypse?' In *What to Think About Machines That Think*, edited by John Brockman, pp. 408–411. New York: Harper Perennial.

Haskins, Caroline. 2019. 'Dozens of Cities Have Secretly Experimented with Predictive Policing Software'. *Motherboard*, February 6. Online. Available: https:// motherboard.vice.com/en_us/article/d3m7jq/dozens-of-cities-have-secretly-experimented-with-predictive-policing-software (accessed March 8, 2019).

Hayles, N. Katherine. 1999. *How We Became Posthuman: Virtual Bodies in Cybernetics, Literature, and Informatics*. Chicago: University of Chicago Press.

Hayles, N. Katherine. 2005. *My Mother Was a Computer: Digital Subjects and Literary Texts*. Chicago: University of Chicago Press.

Helmreich, Stefan. [1998] 2000. *Silicon Second Nature: Culturing Artificial Life in a Digital World*. Los Angeles: University of California Press.

Heesterman, Johannes Cornelis. 1957. *The Ancient Indian Royal Consecration: The Rājasūya Described According to the Yajus Texts and Annotated*. The Hague: Mouton and Co.

Hefner, Philip. 2009. 'The Animal that Aspires To Be an Angel: The Challenge of Transhumanism'. *Dialog: A Journal of Theology* 48 (2): 164–173.

Heilweil, Rebecca. 2020. 'Elon Musk is One Step Closer to Connecting a Computer to Your Brain'. *Vox*, August 28. Online. Available: https://www.vox.com/recode/ 2020/8/28/21404802/elon-musk-neuralink-brain-machine-interface-research (accessed August 30, 2020).

Heinlein, Robert. 1958. *Methuselah's Children*. New. York: New American Library.

Heinlein, Robert. 1973. *Time Enough for Love: The Lives of Lazarus Long*. New York: G.P. Putnam's Sons.

Hejazi, Sara. 2019. '"Humankind. The Best of Molds"—Islam Confronting Transhumanism'. *Sophia* 58(4): 677–688.

Helmreich, Stefan. [1998] 2000. *Silicon Second Nature: Culturing Artificial Life in a Digital World*. Los Angeles: University of California Press.

Herzfeld, Noreen. 2002. 'Creating in Our Own Image: Artificial Intelligence and the Image of God'. *Zygon: Journal of Religion and Science* 37(4): 303–316.

Herzfeld, Noreen. 2002. 'Cybernetic Immortality Versus Christian Resurrection'. In *Resurrection: Theological and Scientific Arguments*, edited by Ted Peters, Robert John Russell, and Michael Welker, pp. 192–201. Grand Rapids, MI: William B. Eerdmans.

Herzfeld, Noreen. 2009. *Technology and Religion: Remaining Human in a Co-created World*. West Conshohocken, PA: Templeton.

Hillis, Daniel W. 2001. 'A Time of Transition/The Human Connection'. In *True Names and the Opening of the Cyberspace Frontier*, edited by James Frenkel, pp. 27–32. New York: Tor.

Hillis, Daniel W. 2019. 'The First Machine Intelligences'. In *Possible Minds: Twenty-Five Ways of Looking at AI*, edited by John Brockman, pp. 170–180. New York: Penguin.

Hochman, David. 2016. 'Reinvent Yourself'. *Playboy*, April 19. Online. Available: http:// www.kurzweilai.net/playboy-reinvent-yourself-the-playboy-interview (accessed December 29, 2018).

Honglardarom, Soraj. 2002. 'The Web of Time and the Dilemma of Globalization'. *The Information Society* 18(4): 241–249.

Honglardarom, Soraj. 2009. 'Nanotechnology, Development and Buddhist Values'. *Nanoethics* 3(2): 97–107.

Horsley, Richard A. 1993. *Jesus and the Spiral of Violence: Popular Jewish Resistance in Roman Palestine*. Minneapolis: Fortress Press.

Husain, Amir. 2017. *The Sentient Machine: The Coming Age of Artificial Intelligence*. New York: Scribner.

Hutchings, Kimberly. 2019. 'Decolonizing Global Ethics: Thinking with the Pluriverse'. *Ethics & International Affairs* 33(2): 115–125.

Huxley, Aldous. [1932] 2014. *Brave New World* (Kindle edition). New York: HarperCollins.

Huxley, Julian. 1955. 'Morals without Religion'. *The Indian Rationalist* 3(8): 114–115.

Huxley, Julian. [1927] 1957. *Religion Without Revelation*. New York: Harper and Row.

Huxley, Julian. 1957. *New Bottles for New Wine*. New York: Harper and Brothers.

Huxley, Julian. 1963. 'The Future of Man—Evolutionary Aspects'. In *Man and His Future*, edited by Gordon Wolstenholme, pp. 1–22. Boston: Little, Brown and Company.

International Humanist Congress. 1953. '"The Humanist Appeal." Republication of a resolution by the International Humanist Congress to serve as preamble to the by-laws of a future World Union'. *The Indian Rationalist* 1(6): 48.

Isenberg, Max. 1953. 'Call it Rationalism'. *The Indian Rationalist* 1(12): 89.

Jackelén, Antje. 2002. 'The Image of God as *Techno Sapiens*'. *Zygon: Journal of Religion and Science* 37 (2): 289–302.

Jackson, Roy. 2020. *Muslim and Supermuslim: The Quest for the Perfect Being and Beyond*. Cham, Switzerland: PalgraveMacmillan.

Jacobsen, Knut A. 2018. 'Pilgrimage Rituals and Technological Change: Alternations in the *Shraddha* Ritual at Kapilashram in the town of Siddhpur in Gujarat'. In *Religion and Technology in India: Spaces, Practices and Authorities*, edited by Knut A. Jacobsen and Kristina Myrvold, pp. 130–145. New York: Routledge.

Jacobsen, Knut A. and Kristina Myrvold (eds.). 2018. *Religion and Technology in India: Spaces, Practices and Authorities*. New York: Routledge.

Jain, Akshai. 2018. 'A Kerala Botanist's Affair With an Unlikely 17th Century Book'. *The Wire*, September 30. Online. Available: https://thewire.in/the-sciences/hortus-malabaricus-van-rheede-ks-manilal-botany-itty-achuden (accessed November 27, 2018).

Jamison, Stephanie. 1996. *Sacrificed Wife/Sacrificer's Wife: Women, Ritual, and Hospitality in Ancient India*. New York: Oxford University Press.

Johnson, Phillip E. 1990. 'Evolution as Dogma: The Establishment of Naturalism'. *First Things* 6(October): 15–22.

Jonze, Spike. 2013. *Her*. Los Angeles: Warner Bros.

Joint Economic Committee (U.S. Congress). 2007. 'Nanotechnology: The Future Is Coming Sooner than You Think'. Washington, D.C.: Joint Economic Committee.

Joy, Bill. 2000. 'Why the Future Doesn't Need Us'. *Wired* 8(4). Online. Available: https://www.wired.com/2000/04/joy-2/ (accessed April 2, 2019).

Jung, Daekyung. 2019. 'Transhumanism and the Theology of *Xiang*: Deconstructing Transhumanism's Self-Centered Epistemology and Retrieving a Communal Sense of Being'. *Theology and Science* 17(4): 524–538.

Kaku, Michio. 1997. *Visions: How Science Will Revolutionize the 21st Century*. New York: Anchor.

Kaku, Michio. 2018. *The Future of Humanity: Terraforming Mars, Interstellar Travel, Immortality and Our Destiny Beyond Earth*. New Delhi: Allen Lane.

Kamble, V.B. 2001. 'Clones—It's Human Beings Now'. *Dream 2047* 3(11): 35.

Kamble, V.B. 'Making Science More Accessible and Less Frightening'. *Dream 2047* 3(4): 23.

Kapil, Raj. 2013. 'Beyond Postcolonialism ... and Postpositivism: Circulation and the Global History of Science'. *Isis* 104(2): 337–347.

Kaplan, Jerry. 2016. *Artificial Intelligence: What Everyone Needs to Know*. New York: Oxford University Press.

Karmakar, R.D. 1953. 'The Pariplava (Revolving Cycle of Legends) at the Asvamedha'. *Annals of the Bhandarkar Oriental Research Institute* 33(1): 26–40.

Kass, Leon. 2003. 'Letter of Transmittal to the President'. In *Beyond Therapy: Biotechnology and the Pursuit of Happiness—A Report of the President's Council on Bioethics*, pp. xv–xiix. Washington, D.C.: The President's Council on Bioethics

Kaunda, Chammah Judex. 2020. 'Bemba Mystico-Relationality and the Possibility of Artificial General Intelligence (AGI) Participation in *Imago Dei*'. *Zygon* 55(2): 327–343.

Kelly, Kevin. 2017. 'The Myth of a Superhuman AI'. *Wired* 25(4). Online. Available: https://www.wired.com/2017/04/the-myth-of-a-superhuman-ai/   (accessed April 3, 2019).

Khened, Shivaprasad M. 2006. 'Gordon Moore, His Law, and Integrated Circuits'. *Dream 2047* 9(1): 35–31 (reverse pagination).

Khilnani, Sunil. [1997] 1999. *The Idea of India*. New York: Farrar, Straus and Giroux.

Kimura, Takeshi. 2018. 'Masahiro Mori's Buddhist Philosophy of Robot'. *Paladyn, Journal of Behavioral Robotics* 9(1): 72–81.

Klostermaier, Klaus K. 1989. *A Survey of Hinduism*. Albany: SUNY Press.

Knapp, Alex. 2012. 'Ray Kurweil's Predictions for 2009 Were Mostly Inaccurate'. *Forbes*, March 20. Online. Available: https://www.forbes.com/sites/alexknapp/2012/03/20/ray-kurzweils-predictions-for-2009-were-mostly-inaccurate/#45fdeef23f9a (accessed November 15, 2018).

Knowledge in Civil Society. 2011. *Knowledge Swaraj: An Indian Manifesto on Science and Technology*. Secunderabad: Knowledge in Civil Society Forum. Online. Available: http://kicsforum.net/kics/kicsmatters/Knowledge-swaraj-an-Indian-S&T-manifesto.pdf

Kohli, Deepak. 2019. 'Artificial Intelligence: New Dimensions'. *Dream 2047* 21(7): 29–28 (reverse pagination).

Koppers, W. 1939. 'The Mundas and the Sidoli Feast of the Korkoos: On the Traces of the Ancient Aśvamedha'. *The New Review* 10(57): 201–212.

Kostik, Kristin, Lea Fowler, and Christopher Scott. 2019. 'Engineering Eden: Does Earthly Pursuit of Eternal Life Threaten the Future of Religion?' *Theology and Science* 17(2): 209–222.

Kuhn, Thomas. [1962] 1996. *The Structure of Scientific Revolutions*. Chicago: University of Chicago Press.

Kumar, Deepak. 2006. *Information Technology and Social Change: A Study of Digital Divide in India*. New Delhi: Rawat.

Kumar, G.S. 2013. 'Community Space Wins Global Acclaim'. *Times of India*, September 25. 12). Online. Available: http://timesofindia.indiatimes.com/city/bengaluru/community-space-wins-global-acclaim/articleshow/23012415.cms (accessed October 13, 2013; no longer available).

Kurzweil, Ray. 1999. *The Age of Spiritual Machines: When Computers Exceed Human Intelligence*. New York: Viking.

Kurzweil, Ray. 2005. *The Singularity Is Near: When Humans Transcend Biology*. New York: Viking.

Kurzweil, Ray. 2016. 'Ray Kurzweil Talks At the Nobel Prize Events'. *KurzweilAI.com*, January 1. Online. Available: http://www.kurzweilai.net/ray-kurzweil-keynote-and-panel-at-nobel-week-dialog-from-the-nobel-prize (accessed April 4, 2019).

Kurzweil, Ray. 2019. *Danielle: Chronicles of a Superheroine*. Monument, CO: WordFire.

Kushner, David. 2009. 'When Man & Machine Merge'. *Rolling Stone* 1072(2009): 56–61.

Laine, James W. 2010. 'Mind and Mood in the Study of Religion'. *Religion* 40(4): 239–249.

Lanier, Jaron. 2000. 'One Half A Manifesto'. *Edge*, November 10. Online. Available: https://www.edge.org/conversation/jaron_lanier-one-half-a-manifesto (accessed April 2, 2019).

Lanier, Jaron. 2010. *You Are Not a Gadget: A Manifesto*. New York: Knopf.

Latour, Bruno. [1991] 1993. *We Have Never Been Modern*, translated by Catherine Porter. Cambridge, MA: Harvard University Press.

Latour, Bruno. 2005. *Reassembling the Social: An Introduction to Actor-Network Theory*. New York: Oxford University Press.

Law, Narendranath. 1918. 'The Horse-Sacrifice and Its Political Significance'. *Modern Review* 23(6): 634–640.

Lepri, Bruno, Nuria Oliver, Emmanuel Letouzé, Alex Pentland, and Patrick Vinck. 2018. 'Fair, Transparent, and Accountable Algorithmic Decision-making Processes: The Premise, the Proposed Solutions, and the Open Challenges'. *Philosophy and Technology* 31(4): 611–627.

Levy, David. 2006. *Robots Unlimited: Life in a Virtual Age*. Wellesley, MA: A.K. Peters.

Levy, David. 'Some Vedic Rituals and Their Political Significance'. *Modern Review* 23(5): 532–534.

Livingston, Steven and Mathias Risse. 2019. 'The Future Impact of Artificial Intelligence on Humans and Human Rights'. *Ethics & International Affairs* 33(2): 141–158.

Lloyd, Seth. 2019. 'Wrong, But More Relevant than Ever'. In *Possible Minds: Twenty-Five Ways of Looking at AI*, edited by John Brockman, pp. 1–12. New York: Penguin.

Löffler, Diana, Jörn Hurtienne, and Ilona Nord. 2019. 'Blessing Robot BlessU2: A Discursive Design Study to Understand the Implications of Social Robots in Religious Contexts'. *International Journal of Social Robotics*. Advance on-line access.

Lourdusamy, John. 2004. *Science and National Consciousness in Bengal (1870–1930)*. London: Sangam.

Luchesi, Brigitte. 2018. 'Modern Technology and Its Impact on Religious Performances in Rural Himachal Pradesh: Personal Remembrances and Observations'. In *Religion and Technology in India: Spaces, Practices and Authorities*, edited by Knut A. Jacobsen and Kristina Myrvold, pp. 112–129. New York: Routledge.

Luciano, Floridi. 2020. 'Mind the App—Considerations on the Ethical Risks of COVID-19 Apps'. *Philosophy & Technology* 33(2): 167–172.

Luker, Victoria. 1998. 'Millenarianism in India: The Movement of Birsa Munda'. In *Religious Traditions in South Asia: Interaction and Change*, edited by Geoffrey A. Oddie, pp. 51–64. Richmond, Surry: Curzon Press.

Mackenzie, Donald and Judy Wajcman. 1985. 'Introductory Essay'. In *Reading the Social Shaping of Technology*, edited by Donald Mackenzie and Judy Wajcman, pp. 1–49. Milton Keynes: Open University Press. Online. Available: https://eprints.lse.ac.uk/28638/1/Introductory essay (LSERO).pdf

Mādhavānanda (translator). 1950. *Bṛhadāraṇyaka Upaniṣad with the Commentary of Śankarācārya*. Mayavati: Advaita Ashrama.

Mahanti, Subodh. 2004. 'John Burdon Sanderson Haldane: The Ideal of a Polymath'. *Dream 2047* 6(3): 34–30 (reverse pagination).

Maitra, Herambachandra. 1919. *The Modern Review* 26(6): 597–604.

Maitra, Soumya. 2016. 'Demystifying the Human Brain'. *Dream 2047* 18(12): 31–30 and 22 (reverse pagination).

Malamud, Carl. 2018. 'Note on Visit to Sabarmati Ashram'. In *Code Swaraj: Fieldnotes from the Standards Satyagraha*, edited by Carl Malamud and Sam Pitroda, pp. 17–27. Sebastopol: Public.Resource.Org, Inc.

Malik, Udit. 2018. 'Artificial Intelligence – Changing the World Like Never Before!' *Dream 2047* 20(9): 26–25 (reverse pagination).

Manu. 1957. 'Laws of Manu'. In *A Sourcebook of Indian Philosophy*, edited by Sarvepalli Radhakrishnan and Charles A. Moore, pp. 172–192. Princeton, NJ: Princeton University Press.

Martin, George. 1971. 'Brief Proposal on Immortality: An Interim Solution'. *Perspectives in Biology and Medicine* 14(2): 339–40.

Marty, Martin E. 2020. 'Still Searching for a Usable Future'. *Sightings*, February 10. Online. Available: https://divinity.uchicago.edu/sightings/articles/still-searching-usable-future (accessed February 11, 2020).

Mauss, Marcel. [1950] 2000. *The Gift: The Form and Reason for Exchange in Archaic Societies*. New York: W.W. Norton.

Mayson, Sandra G. 2019. 'Bias In, Bias Out'. *Yale Law Journal* 128(8): 2218–2300.

McCutcheon, Russell. 2001. *Critics Not Caretakers: Redescribing the Public Study of Religion*. Albany: State University of New York Press.

McGrath, James and Ankur Gupta. 2018. 'Writing a Moral Code: Algorithms for Ethical Reasoning by Humans and Machines'. *Religions* 9(8): 1–19.

McLaughlin, William G. 1959. *Modern Revivalism: Charles Grandison Finney to Billy Graham*. New York: Charles Scribner's Sons.

McMillan, Robert. 2015. 'AI Has Arrived, and That Really Worries the World's Brightest Minds'. *Wired* 23(1). Online. Available: https://www.wired.com/2015/01/ai-arrived-really-worries-worlds-brightest-minds/ (accessed April 4, 2019).

McOuat, Gordon. 2017. 'J.B.S. Haldane's Passage to India: Reconfiguring Science'. *Journal of Genetics* 96(5): 845–852.

Meeks, Wayne. 2000. 'Apocalyptic Discourse and Strategies of Goodness'. *The Journal of Religion* 80(3): 461–475.

Mehra, Gargi. 2018. 'The Society of Flower-Pickers'. In *The Best Asian Speculative Fiction*, edited by Rajat Chaudhuri, pp. 74–89. Singapore: Kitaab.

Menon, Anil. 2009. *The Beast with Nine Billion Feet*. New Delhi: Zubaan. Unpaginated Kindle edition.

Menon, Anil. 2019. 'Shit Flower'. In *The Gollancz Book of South Asian Science Fiction*, edited by Tarun K. Saint and Manjula Padmanabhan, pp. 51–73. Gurugram: Hachette India.

Mercer, Calvin and Tracy J. Trothen (eds.). 2014. *Religion and Transhumanism: The Unknown Future of Human Enhancement*. Santa Barbara: Praeger.

Mercer, Calvin and Derek F. Maher (eds.) 2014. *Transhumanism and the Body: The World Religions Speak*. New York: Palgrave.

Metchnikoff, Élie. 1905. *The Nature of Man: Studies in Optimistic Philosophy*, translated by P. Chalmers Mitchell. New York: G.P. Putnam's Sons.

Metzinger, Thomas. 2019. 'EU Guidelines: Ethics Washing Made in Europe'. *Der Tagesspiegel*, April 8. Online. Available: https://m.tagesspiegel.de/politik/eu-guidelines-ethics-washing-made-in-europe/24195496.html (accessed May 13, 2019).

Midgley, Mary. 1992. *Science as Salvation*. New York: Routledge.

Minsky, Marvin. 1994. 'Will Robots Inherit the Earth?' *Scientific American*, October 1. Online. Available: web.media.mit.edu/~minsky/papers/sciam.inherit.html (accessed April 1, 2019).

Mishra, Krishna Kumar. 2015. 'Nanoscience and Its Applications'. *Dream 2047* 17(6): 34–32 (reverse pagination).

Misra, Kinkini Dasgupta. 2002. 'Shaping the World Atom by Atom'. *Dream 2047* 4(10): 21–20 (reverse pagination).

Misra, Maria. [2007] 2008. *Vishnu's Crowded Temple: India Since the Great Rebellion*. New Haven, CT: Yale University Press.

Mohamed, Shakir, Marie-Therese Png, and William Isaac. 2020. 'Decolonial AI: Decolonial Theory as Sociotechnical Foresight in Artificial Intelligence'. *Philosophy & Technology* 33(4): 659–684.

Mohanty, Ashutosh. 2016. 'Nanotechnology in Environmental Remediation'. *Dream 2047* 18(10): 25.

Mookerjee, Dhirendra Nath. 1945. 'The Krta Era'. *Journal of Indian History* 24(3): 104–109.

Moravec, Hans. [1976] 1978. 'Today's Computers, Intelligent Machines and Our Future'. *Analog* 99(2): 59–84. Online. Available: https://frc.ri.cmu.edu/~hpm/project.archive/general.articles/1978/analog.1978.html (accessed March 27, 2019).

Moravec, Hans. 1988. *Mind Children: The Future of Robot and Human Intelligence*. Cambridge, MA: Harvard University Press.

Moravec, Hans. 1992. 'Pigs in Cyberspace'. In *Thinking Robots, An Aware Internet, and Cyberpunk Librarians: The 1992 LITA President's Program*, edited by R. Bruce Miller and Milton T. Wolf, pp. 15–21. Chicago: Library and Information Technology Association.

Moravec, Hans. 1999. *Robot: The Future of Machine and Human Intelligence*. New York: Oxford University Press.

More, Max. [1988] 2003. 'Principles of Extropy: Version 3.11'. Extropy Institute. Online: www.extropy.org/principles (accessed October 4, 2008; no longer active).

More, Max. 2013. 'The Philosophy of Transhumanism'. In *The Transhumanist Reader: Classical and Contemporary Essays on the Science, Technology, and Philosophy of the Human Future*, edited by Max More and Natasha Vita-More, pp. 3–17. Malden, MA: Wiley-Blackwell.

Morris, Henry and John C. Whitcomb. *The Genesis Flood: The Biblical Record and Its Scientific Implications*. Philadelphia: Presbyterian and Reformed Publishing.

Mubin, Omar, Kewal Wadibhasme, Philipp Jordan, and Mohammad Obaid. 2019. 'Reflecting on the Presence of Science Fiction Robots in Computing Literature'. *ACM Transactions in Human-Robot Interaction* 8(1): Article 5.

Murthy, M.S.S. 2017. 'New Tools for Gene Therapy'. *Dream 2047* 19(5): 34–32 (reverse pagination).

Munshi, Nupur. 2016. 'Turing Church Manifests Durga's Symbolism of Good over Evil'. *India Future Society webpage*, October 25, 2015. Online. Available: http://indiafuturesociety.org/turing-church-manifests-durgas-symbolism-of-good-over-evil/ (accessed January 19, 2016).

Munshi, Nupur. 2017. 'Building Gods'. *Turing Church webpage*, August 29. Online. Available: https://turingchurch.net/building-gods-f31aa9171676 (accessed April 23, 2019).

Musk, Elon. 2014. Twitter post, August 2. Online. Available: https://twitter.com/elonmusk/status/495759307346952192?lang=en (accessed April 2, 2019).

Musk, Elon. 2014. *Twitter post*, August 3. Online. Available: https://twitter.com/elonmusk/status/496012177103663104?lang=en (accessed April 2, 2019).

Nagaraj, Nithin. 2020. '"AI: From Turing to Sophia." Presented at the "Facets of AI"'. Workshop hosted by the National Institute of Advanced Studies (July 15). Bangalore, India. Online. Available: https://www.youtube.com/watch?v=R4ylT5_Y7Es&feature=youtu.be.

Nagendra, Harini. 2016. *Nature in the City: Bengaluru in the Past, Present, and Future*. Delhi: Oxford University Press.

Nair, Janaki. 2005. *The Promise of the Metropolis: Bangalore's Twentieth Century*. New Delhi: Oxford University Press.

Nanda, Meera. 2003. *Prophets Facing Backwards: Postmodern Critiques of Science and Hindu Nationalism in India*. New Brunswick: Rutgers University Press.

Nandy, Ashis. [1980] 2012. *Alternative Sciences*. Reprinted in *Return from Exile*. New York: Oxford University Press.

Nandy, Ashis. [1983] 2012. *The Intimate Enemy: Loss and Recovery of Self under Colonialism*. New York: Oxford University Press.

Narayan, Kirin and Kenneth M. George. 2017. 'Tools and World-Making in the Worship of Vishwakarma'. *South Asian History and Culture* 8(4): 478–492.

Narayan, Kirin and Kenneth M. George. 2018. 'Vishwakarma: God of Technology'. In *Religion and Technology in India: Spaces, Practices and Authorities*, edited by Knut A. Jacobsen and Kristina Myrvold, pp. 8–24. New York: Routledge.

Narlikar, Jayant. 2003. *The Scientific Edge: The Indian Scientist from Vedic to Modern Times*. New Delhi: Penguin.

Nath, Vijay. 2001. 'From "Brahmanism" to "Hinduism": Negotiating the Myth of the Great Tradition'. *Social Scientist* 29(3/4): 19–51.

National Public Radio. [1999] 2018. 'The Science in Science Fiction'. *NPR.org*, October 22. Online. Available: https://www.npr.org/2018/10/22/1067220/the-science-in-science-fiction (accessed April 25, 2019).

Nehru, Jawaharlal. [1946] 2010. *The Discovery of India*. New York: Penguin.

Nehru, Jawaharlal. 1988. 'The Need for a Spirit of Service: Speech at the Inaugural Meeting of the All-India Scientific Workers' Association at New Delhi on January 7, 1947'. In *Jawaharlal Nehru on Science and Society: A Collection of His Writings and Speeches*, edited by Baldev Singh, pp. 43–44. New Delhi: Nehru Memorial Museum and Library.

Nehru, Jawaharlal. 'The Spirit of Science: Speech on the Occasion of the Opening of the Central Fuel Research Institute, Jealgora, on April 22, 1950'. In *Jawaharlal Nehru on Science and Society: A Collection of His Writings and Speeches*, edited by Baldev Singh, pp. 77–79. New Delhi: Nehru Memorial Museum and Library.

Newell, Allen. 1990. 'Fairy Tales'. In *The Age of Intelligent Machines*, edited by Raymond Kurzweil, pp. 420–423. Cambridge, MA: The MIT Press.

NITI Aayog. 2018. 'National Strategy for Artificial Intelligence'. *NITI Aayog*. Online. Available: https://www.niti.gov.in/writereaddata/files/document_publication/NationalStrategy-for-AI-Discussion-Paper.pdf (accessed August 18, 2019).

Noble, David. 1999. *The Religion of Technology: The Divinity of Man and the Spirit of Invention*. New York: Penguin.

Nourbakhsh, Illah Reza. 2013. *Robot Futures*. Cambridge, MA: The MIT Press.

Nourbakhsh, Illah Reza and Jennifer Keating. 2019. *AI and Humanity*. Cambridge, MA: The MIT Press.

Nye, David E. 2003. *America as Second Creation: Technology and Narratives of a New Beginning*. Cambridge, MA: MIT University Press.

O'Reilly, Tim. 2015. 'What if We're the Microbiome of Silicon AI?' In *What To Think about Machines That Think*, edited by John Brockman, pp. 153–155. New York: Harper Perennial.

ÓeÉigertaigh, Seàn, Jess Whittlestone, Yang Liu, Yi Zeng, and Zhe Liu. 2020. 'Overcoming Barriers to Cross-cultural Cooperation in AI Ethics and Governance'. *Philosophy & Technology* 33(4): 571–593.

Ornella, Alexander. 2015. 'Towards a 'Circuit of Technological Imaginaries': A Theoretical Approach'. In *Religion in Cultural Imaginary: Explorations in Visual and Material Practices*, edited by Daria Pezzoli-Olgiati, pp. 303–332. Baden-Baden: Nomos.

Pablo. 2005. 'Transhumanism and FM Esfandiary'. *Institute for Ethics and Emerging Technologies website*, January 24. Online. Available: https://ieet.org/cybdem/2005/01/transhumanism-and-fm-esfandiary.html (accessed October 5, 2020).

Pace, Ben. 2019. 'Debate on Instrumental Convergence between LeCun, Russell, BEngio, Zador, and More'. *AI Alignment Forum*, October 4. Online. Available: https://www.alignmentforum.org/posts/WxW6Gc6f2z3mzmqKs/debate-on-instrumental-convergence (accessed October 7, 2019).

Pagallo, Ugo. 2018. 'Algo-Rhythms and the Beat of the Legal Drum'. *Philosophy & Technology* 31(4): 507–524.

Palmås, Karl. 2011. 'Predicting What You'll Do Tomorrow: Panspectric Surveillance and the Contemporary Corporation'. *Surveillance & Society* 8(3): 338–354.

Pandey, Poonam and Aviram Sharma. 2020. 'Swinging between responsibility and rationality – Science policy and technology visions in India'. In *Diece neutrale Normativität der Technikfolgenabschätzung konzeptionelle Auseinandersetzung und praktischer Umgang*, edited by Linda Nierling and Helge Torgersen, pp. 155–174. Baden-Baden: Nomos.

Paramahansa, Yogananda. [1946] 2011. *Autobiography of a Yogi*. Kolkata: Yogoda Satsanga Society of India.

Paranjape, Makarand. 2008. 'Science, Spirituality and Modernity in India'. In *Science, Spirituality and the Modernization of India*, edited by Makarand Paranjape, pp. 3–14. New Delhi: Anthem.

Parthasarathi, Prasannan. 2001. *The Transition to a Colonial Economy: Weavers, Merchants and Kings in South India 1720–1800*. Cambridge: University of Cambridge Press.

Parthasarathi, Prasannan. 2011. *Why Europe Grew Rich and Asia Did Not: Global Economic Divergence, 1600–1850*. New York: Cambridge University Press.

Perera, Sasanka. 2015. *Debating the Ancient and Present: A Conversation with Romila Thapar*. Delhi: Aakar.

Pein, Corey. 2016. 'Everybody Freeze!: The Extropians Want Your Body'. *The Baffler* 30: 84–101.

Perkowitz, Sydney. 2004. *Digital People: From Bionic Humans to Androids*. Washington, D.C.: Joseph Henry.

Pesce, Mark. 2001. 'True Magic'. In *True Names and the Opening of the Cyberspace Frontier*, edited by James Frenkel, pp. 221–238. New York: Tor.

Peters, Jay. 2020. 'IBM Will No Longer Offer, Develop, or Research Facial Recognition Technology'. *The Verge*, June 8. Online. Available: https://www.theverge.com/2020/6/8/21284683/ibm-no-longer-general-purpose-facial-recognition-analysis-software (accessed June 8, 2020).

Pfister, Wally. 2014. *Transcendence*. Los Angeles: Warner Bros.

Phondke, Bal (ed.). 1993. *It Happened Tomorrow*. New Delhi: National Book Trust.

Pinch, Trevor J. and Wiebe E. Bijker. 1984. 'The Social Construction of Facts and Artefacts: Or How the Sociology of Science and the Sociology of Technology might Benefit Each Other'. *Social Studies of Science* 14(3): 399–441.

Pitroda, Sam. 2018. 'Access to Knowledge in America and India'. In *Code Swaraj: Fieldnotes from the Standards Satyagraha*, edited by Carl Malamud and Sam Pitroda, pp. 31–37. Sebastopol: Public.Resource.Org, Inc.

Pitroda, Sam. 'Right to Information, Right to Knoweldge'. In *Code Swaraj: Fieldnotes from the Standards Satyagraha*, edited by Carl Malamud and Sam Pitroda, pp. 63–70. Sebastopol: Public.Resource.Org, Inc.

Pohl, Frederik. 1975. 'The Tunnel Under the World'. In *The Best of Frederik Pohl*, edited by Lester del Rey, pp. 8–35. Garden City, N.Y.: Nelson Doubleday.

Porter, Jon. 2019. 'Federal Study of Top Facial Recognition Algorithms Finds 'Empirical Evidence' of Bias'. *The Verge*, December 20. Online. Available: https://www.theverge.com/2019/12/20/21031255/facial-recognition-algorithm-bias-gender-race-age-federal-nest-investigation-analysis-amazon (accessed July 2, 2020).

Prakash, Gyan. 1999. *Another Reason: Science and the Imagination of Modern India*. Princeton, NJ: Princeton University Press.

Prakash, Vishwa. Unknown publication date. *Life and Teachings of Swami Dayanand.* Allahabad: Kala.

Prasad, Archana. 2010. 'Enlightened Singularity'. *ArchanaPrasad.com*, October 25. Online. Available: http://www.archanaprasad.com/2010/10/25/enlightened-singularity/ (accessed May 22, 2019).

Prasad, Archana. 2015. Email correspondence with the author. July 27, 2015.

Prasad, Jayan. 2019. Email correspondence with the author. December 6, 2019.

Prasad, Jayan. 2019. Email correspondence with the author. December 26, 2019.

Prasad, Jayan. 2019. 'Singularity Cafe Chennai Part 3: Why the Singularity Is Good and How to Achieve It'. Online. Available: https://www.youtube.com/watch?v=ONnNmzoVfqo (accessed December 2, 2019).

President's Council on Bioethics. 2003. *Beyond Therapy: Biotechnology and the Pursuit of Happiness.* Washington, D.C.: President's Council on Bioethics.

Prisco, Giulio. [2004] 2007. 'Engineering Transcendence'. *COSMI2LE*, December 1. Online. Available: http://cosmi2le.com//index.php/site/more/engineering_transcendence (accessed April 26, 2007; site no longer active).

Prisco, Giulio. 2013. 'Transcendent Engineering'. In *The Transhumanist Reader*, edited by Max More and Natasha Vita-More, pp. 234–240. Malden, MA: Wiley-Blackwell.

Prisco, Giulio. 2018. *Tales of the Turing Church.* Middletown, DE: Independently published.

Prisco, Giulio. 2020. 'Pigs in Cyberspace: Neuralink and the Noosphere'. *TuringChurch.net*, September 1. Online. Available: https://turingchurch.net/pigs-in-cyberspace-neuralink-and-the-noosphere-d783f53f699 (accessed September 4, 2020).

Quack, Johannes. 2012. *Disenchanting India: Organized Rationalism and Criticism of Religion in India.* New York: Oxford University Press.

Quote Investigator. 2012. 'The Future Has Arrived—It's Just Not Evenly Distributed Yet'. *Quoteinvestigator.com*. Online. Available: https://quoteinvestigator.com/2012/01/24/future-has-arrived/ (accessed April 25, 2019).

Radhakrishnan, Sarvepalli. 1929. *Kalki, or, The Future of Civilization.* London: Kegan Paul, Trench, Trubner & Co.

Rahwan, Iyad, Manuel Cebrian, Nick Obradovich, Josh Bongard, Jean-François Bonnefon, Cynthia Breaseal, Jacob W. Crandall, Nicholas A. Christakis, Iain D. Couzin, Matthew O. Jackson, Nicholas R. Jennings, Ece Kamar, Isabel M. Kloumann, Hugo Larochelle, David Lazer, Richard McElreath, Alan Mislove, David C. Parkes, Alex 'Sandy' Pentland, Margaret E. Roberts, Azim Shariff, Joshua B. Tenenbaum, and Michael Wellman. 2019. 'Machine Behavior'. *Nature* 568(7753): 477–486.

Raina, Dhruv. [2003] 2010. *Images and Contexts: The Historiography of Science and Modernity in India.* New Delhi: Oxford University Press.

Raina, Dhruv and S. Irfan Habib. 1996. 'The Moral Legitimation of Modern Science: Bhadralok Reflections on Theories of Evolution'. *Social Studies of Science* 26(1): 9–42.

Rajagopal, Krishnadas. 2019. 'Order on Surveillance Meant to Protect Privacy, Govt. Tells SC'. *The Hindu*, March 1. Online. Available: https://www.thehindu.com/news/national/order-on-surveillance-meant-to-protect-privacy-govt-tells-sc/article26412542.ece (accessed March 8, 2019).

Raju, Raghuram. 2008. 'Sri Aurobindo and Krishnachandra Bhattacharya on Science and Spirituality'. In *Science, Spirituality and the Modernization of India*, edited by Makarand Paranjape, pp. 96–114. New Delhi: Anthem.

Ramachandran, T.N. 1953. 'Aśvameda Site Near Kalsi'. *The Journal of Oriental Research, Madras* 22(1): 1–31.

Ramakrishnan, Venki. 2019. 'Will Computers Become Our Overlords?' In *Possible Minds: Twenty-Five Ways of Looking at AI*, edited by John Brockman, pp. 181–191. New York: Penguin.

Raman, Varadaraja V. 2011. *Indic Visions in an Age of Science*. New York: Metanexus.

Ramanathan, S. 1952. 'Humanism and Rationalism'. *The Indian Rationalist* 1(6): 41–42.

Ramanathan, S. 1954. 'Julian Huxley'. *The Indian Rationalist* 2(5): 53–54.

Ramanathan, S. 1957. 'Welcome Haldane'. *The Indian Rationalist* 5(6/7): 5.

Rambelli, Fabio. 2018. 'Dharma Devices, Non-Hermeneutical Libraries, and Robot-Monks: Prayer Machines in Japanese Buddhism'. *Journal of Asian Humanities at Kyushu University* 3: 57–75.

Rapier, Graham. 2019. '"If You Can't Beth Them Join Them": Elon Musk Says Our Best Hope for Competing with AI is Becoming Better Cyborgs'. *Business Insider*, August 29. Online. Available. https://www.businessinsider.com/elon-musk-humans-must-become-cyborgs-to-compete-with-ai-2019-8 (accessed August 30, 2020).

Ray, Praphulla Chandra. [1903] 1904. *A History of Hindu Chemistry*, Volume 1. Calcutta: Chuckervertty, Chatterjee, & Co.

Ray, Praphulla Chandra. [1909] 1925. *A History of Hindu Chemistry*, Volume 2. Calcutta: Chuckervertty, Chatterjee, & Co.

Rees, Martin. 2015. 'Organic Intelligence Has No Long-Term Future'. In *What To Think About Machines That Think*, edited by John Brockman, pp. 9–11. New York: Harper Perennial.

Rennie, John. 2010. 'Ray Kurzweil's Slippery Futurism'. *IEEE Spectrum*, November 29. Online. Available: https://spectrum.ieee.org/computing/software/ray-kurzweils-slippery-futurism/1 (accessed April 2, 2019).

Rennie, John. 2011. 'The Immortal Ambitions of Ray Kurzweil: A Review of *Transcendent Man*'. *Scientific American*, February 15. Online. Available: https://www.scientificamerican.com/article/the-immortal-ambitions-of-ray-kurzweil/ (accessed April 2, 2019).

Rheingold, Howard. 1991. *Virtual Reality: The Revolutionary Technology of Computer-Generated Artificial Worlds—and How It Promises to Transform Society*. New York: Touchstone.

Robertson, Jennifer. 2018. *Robo Sapiens Japanicus: Robots, Gender, Family, and the Japanese Nation*. Oakland: University of California Press.

Robinett, Warren. 2002. 'The Consequences of Fully Understanding the Brain'. In *Converging Technologies for Improving Human Performance: Nanotechnology, Biotechnology, Information Technology and Cognitive Science*, edited by Mihail C. Roco and William Sims Bainbridge, pp. 166–169. Arlington, VA: U.S. National Science Foundation and Department of Commerce.

Robison, Peter. 2019. 'Boeing's 737 Max Software Outsourced to $9-an-Hour Engineers'. *Bloomberg News*, June 28. Online. Available: https://www.bloomberg.com/news/articles/2019-06-28/boeing-s-737-max-software-outsourced-to-9-an-hour-engineers (accessed August 13, 2019).

Robitzsky, Dan. 2019. 'Scientists Say New Quantum Material Could "Download Your Brain"'. *Futurism.com*, April 11. Online. Available: https://futurism.com/the-byte/scientists-quantum-material-download-brain (accessed April 20, 2019).

Roff, Heather M. 2019. 'Artificial Intelligence: Power to the People'. *Ethics & International Affairs* 33(2): 127–140.

Rosedale, Philip. 2019. *Twitter post*, March 31. Online. Available: https://twitter.com/philiprosedale/status/1112553796167557121 (accessed April 5, 2019).

Roy, Anjali. 2008. 'Faith Outside the Lab'. In *Science, Spirituality and the Modernization of India*, edited by Makarand Paranjape, pp. 229–237. New Delhi: Anthem.

Rozin, Paul and Edward B. Royzman. 2001. 'Negativity Bias, Negativity Dominance, and Contagion'. *Personality and Social Psychology Review* 5(4): 296–320.

Russell, D.S. 1964. *The Method and Message of Jewish Apocalypticism: 200 BC—100 AD*. Philadelphia: Westminster Press.

Šabanović, Selma. 2014. 'Inventing Japan's 'Robotics Culture': The Repeated Assembly of Science, Technology, and Culture in Social Robotics'. *Social Studies of Science* 44(3): 342–367.

Said, Edward. 1978. *Orientalism*. New York: Pantheon.

Saint, Tarun K. 2019. 'Introduction'. In *The Gollancz Book of South Asian Science Fiction*, edited by Tarun K. Saint and Manjula Padmanabhan, pp. ix–xli. Gurugram: Hachette India.

Salwi, Dilip M. 2002. 'National Centre for Software Technology: Focusing on Relevant Research'. *Dream 2047* 4(8): 22.

Sangupta, Hindol. 2018. 'Changing Hindutva by Technology: A Case Study of Hindutva Abhiyan and the Use of Social Media'. In *Religion and Technology in India: Spaces, Practices and Authorities*, edited by Knut A. Jacobsen and Kristina Myrvold, pp. 146–163. New York: Routledge.

Sarkar, Sumit. 1992. '"Kaliyuga," "Chakri" and "Bhakti": Ramakrishna and His Times'. *Economic and Political Weekly* 27(29): 1543–1559 and 1561–1566.

Sarkar, Sumit. 1997. *Writing Social History*. New Delhi: Oxford University Press.

Sarukkai, Sundar. 2008. 'Culture of Technology and ICTs'. In *ICTs and Indian Social Change: Diffusion, Poverty, Governance*, edited by Ashwani Saith, M. Vijayabaskar, and V. Gayathri, pp. 34–58. New Delhi: Sage.

Sathaye, Adheesh A. 2015. *Crossing the Lines of Caste: Viśvāmitra and the Construction of Brahmin Power in Hindu Mythology*. New York: Oxford University Press.

Satyamoorthy, Prateeksha. 2018. 'The Dawn'. In *An Anthology of Indian Sci-Fi Stories*, produced by the Bangalore Sci-Fi Club. Singapore: Amazon Asia-Pacific Holdings. Kindle publication.

Savarkar, V.D. [1923] 1969. *Hindutva: Who Is a Hindu?* Bombay: Veer Savarkar Prakashan and Bhave Ltd.

Schaffer, Simon. 2002. 'The Devices of Iconoclasm'. In *Iconoclash: Beyond the Image Wars in Science, Religion, and Art*, edited by Bruno Latour and Peter Weibel, pp. 498–515. Cambridge, MA: MIT University Press and ZKM Center for Art and Media.

Schmidt, Benjamin. 2015. *Inventing Exoticism: Geography, Globalism, and Europe's Early Modern World*. Philadelphia: University of Pennsylvania Press.

Schoepflin, Rennie B. 2000. 'Apocalypticism in an Age of Science'. In *The Encyclopedia of Apocalypticism*, edited by Bernard McGinn, John J. Collins, and Stephen J. Stein,

Vol. 3, *Apocalypticism in the Modern World and the Contemporary Age*, pp. 427–441. New York: Continuum Press.

Schopen, Gregory. 1991. 'Archeology and Protestant Presuppositions in the Study of Indian Buddhism'. *History of Religions* 31(1): 1–23.

Schussler, Aura-Elena. 2019. 'Transhumanism as a New Techno-Religion and Personal Development: In the Framework of a Future Technological Spirituality'. *Journal for the Study of Religions and Ideologies* 18(53): 92–106.

Seal, Brajendranath. 1915. *The Positive Sciences of the Ancient Hindus*. London: Longmans, Green and Co.

Searle, Joshua T. 2012. 'The Future of Millennial Studies and the Hermeneutics of Hope: A Theological Reflection'. In *Beyond the End: The Future of Millennial Studies*, edited by Joshua Searle and Kenneth G.C. Newport, pp. 142–159. Sheffield: Sheffield Phoenix.

Sehgal, Narendar K. (presumed). 1999. 'Seminar on Information Technology in India in the Next Millennium'. *Dream 2047* 1(9): 1.

Sekhsaria, Pankaj. 2019. *Instrumental Lives: An Intimate Biography of an Indian Laboratory*. London: Routledge.

Sekhsaria, Pakaj and Naveen Thayyil. 2019. 'Technology Vision 2035: Visions, Technologies, Democracy and the Citizen of India'. *Economic and Political Weekly* 54(34): 64–69.

Sen, K.C. 1933. 'The Religion of Man'. *The Calcutta Review* 46(2): 227–260.

Sherman, Justin. 2019. 'Digital Authoritarianism and the Threat to Global Democracy'. *Bulletin of the Atomic Scientists*, July 25. Online. Available: https://thebulletin.org/2019/07/digital-authoritarianism-and-the-threat-to-global-democracy (accessed July 26, 2019).

Shrivastava, Rishabh and Preeti Mahajan. 2016. 'Artificial Intelligence Research in India: A Scientometric Analysis'. *Science & Technology Libraries* 35(2): 136–151.

Singer, Milton. 1972. *When a Great Tradition Modernizes: An Anthropological Approach to Indian Civilization*. New York: Praeger.

Singer, Peter. 2009. *Wired for War: The Robotics Revolution and Conflict in the 21st Century*. New York: Penguin.

Singh, Manu Pratap and Ashish Chaturvedi. 2004. 'Dreamer and Thinker: An Artificial Brain'. *Dream 2047* 6(3): 23.

Singh, Vandana. 2019. 'Reunion'. In *The Gollancz Book of South Asian Science Fiction*, edited by Tarun K. Saint and Manjula Padmanabhan, pp. 341–365. Gurugram: Hachette India.

Singler, Beth. 2019. 'Existential Hope and Existential Despair in AI Apocalypticism and Transhumanism'. *Zygon: Journal of Religion and Science* 54(1): 156–176.

Smelters, Gregory S. 1953. 'Materialism'. *The Indian Rationalist* 1(9): 71.

Smelters, Gregory S. 1955. 'Morals without Gods'. *The Indian Rationalist* 3(10): 140.

Smith, Jonathan Z. 1982. *Imagining Religion: From Babylon to Jonestown*. Chicago: University of Chicago Press.

Srinivas, M.N. [1966] 2013. *Social Change in Modern India*. Hyderabad: Orient Blackswan.

Srinivas, Smriti. 2018. 'Highways for Healing: Contemporaneous 'Temples' and Religious Movements in an Indian City'. *Journal of the American Academy of Religion* 86(2): 473–496.

Srinivas, Tulasi. 2018. *The Cow in the Elevator: An Anthropology of Wonder.* Durham, NC: Duke University Press.

Steinhart, Eric. 2008. 'Teilhard de Chardin and Transhumanism'. *Journal of Evolution and Technology* 20(1): 1–22.

Stephenson, Neal. 1992. *Snow Crash.* New York: Bantam.

Stephenson, Neal. 2015. *Seveneves.* New York: William Morrow.

Strozier, Charles B. and Laura Simich. 1991. 'Christian Fundamentalism and Nuclear Threat'. *Political Psychology* 12(1): 81–96.

Subbarayappa, B.V. 1992. *In Pursuit of Excellence: A History of the Indian Institute of Science.* New Delhi: Tata McGraw-Hill.

Subbarayappa, B.V. 2013. *Science in India: A Historical Perspective.* New Delhi: Rupa.

Subramaniam, Banu. 2019. *Holy Science: The Biopolitics of Hindu Nationalism.* New Delhi: Orient BlackSwan.

Subramanian, Samanth. 2019. *A Dominant Character: The Radical Science and Restless Politics of J.B.S. Haldane.* New Delhi: Simon & Schuster.

Subramanian, Samanth. Email correspondence with the author. March 6, 2020.

Subramanian, Samanth. Email correspondence with the author. March 7, 2020.

Subramanyam, Sanjay. 2017. *Europe's India: Words, People, Empires, 1500–1800.* Cambridge, MA: Harvard University Press.

Sunderland, Jabez T. 1932. 'Why Need We Ever Grow Old?' *Modern Review* 51(1): 4–6.

Sur, Atul K. 1939. 'Asvamedha—A Rejoinder'. *Indian Culture* 5(3): 704–706.

Sutton, Deborah R. 2018. 'So Called Caste: S.N. Balanghadara, the Ghent School and the Politics of Grievance'. *Contemporary South Asia* 26(3): 336–349.

Swisher, Kara. 2020. 'Elon Musk: "A.I. Doesn't Need to Hate Us to Destroy Us"'. *The New York Times*, September 28. Online. Available: https://www.nytimes.com/2020/09/28/opinion/sway-kara-swisher-elon-musk.html (accessed November 13, 2020).

Tagore, Rabindrinath. [1925] 2011. 'Pathway to *Mukti*'. In *Indian Philosophy in English: From Renaissance to Independence*, edited by Nalini Bhushan and Jay L. Garfield, pp. 153–164. New York: Oxford University Press.

Tallinn, Jaan. 2019. 'Dissident Messages'. In *Possible Minds: Twenty-Five Ways of Looking at AI*, edited by John Brockman, pp. 88–99. New York: Penguin.

Tagore, Rabindranath. [1919] 2017. 'The Center of Indian Culture'. In *The Complete Works of Rabindranath Tagore.* New Delhi: General Press. Kindle version.

Tamatea, Laurence. 2010. 'Online Buddhist and Christian Responses to Artificial Intelligence'. *Zygon: Journal of Religion and Science* 45(4): 979–1002.

Tasioulas, John. 2019. 'First Steps Towards an Ethics of Robots and Artificial Intelligence'. *Journal of Practical Ethics* 7(1): 49–83.

Tattvabhushan, Sitanath. 1931. 'Dr Haldane on the Immortality of the Individual'. *The Calcutta Review* 38(2): 321–329.

Taylor, Shelley E. 1991. 'Asymmetrical Effects of Positive and Negative Events: The Mobilization-Minimization Hypothesis'. *Psychological Bulletin* 110(1): 67–85.

Team YS. 2013. 'Archana Prasad, Founder & Director, Jaaga'. *HerStory*, October 12. Online. Available: https://yourstory.com/2013/10/archana-prasad-founder-director-jaaga (accessed October 13, 2013).

Technology Information, Forecasting and Assessment Council. 2016. *Technology Vision 2035: Technology Roadmap on Information & Communications Technology.*

New Delhi: Technology Information, Forecasting and Assessment Council, Department of Science and Technology.

Tegmark, Max. 2017. *Life 3.0: Being Human in the Age of Artificial Intelligence.* New York: Knopf.

Tegmark, Max. 2019. 'Let's Aspire to More than Making Ourselves Obsolete'. In *Possible Minds: Twenty-Five Ways of Looking at AI,* edited by John Brockman, pp. 76–87. New York: Penguin.

Thakur, Vaibhav. 2019. *123 Tomorrows: It's Not a Story, It's a Puzzle.* Singapore: Amazon Asia-Pacific Holdings. Kindle publication.

Thomas, Renny. 2016. 'Being Religious, Being Scientific: Science, Religion, and Atheism in Contemporary India'. In *Science and Religion: East and West,* edited by Yiftach Fehige, pp. 140–157. New York: Routledge.

Thomas, Renny. 2017. 'Atheism and Unbelief among Indian Scientists: Towards an Anthropology of Atheism(s)'. *Society and Culture in South Asia* 3(1): 45–67.

Thomas, Renny. 2018. 'Beyond Conflict and Complementarity: Science and Religion in Contemporary India'. *Science, Technology & Society* 23(1): 47–64.

Thomas, Renny. 2020. 'Brahmins as Scientists and Science as Brahmins' Calling: Caste in an Indian Scientific Research Institute'. *Public Understanding of Science* 29(3): 306–318.

Thomas, Renny and Robert M. Geraci. 2018. 'Religious Rites and Scientific Communities: *Ayudha Puja* as 'Culture' at the Indian Institute of Science'. *Zygon: Journal of Religion and Science* 53(1): 95–122.

Thweatt-Bates, Jeanine. 2013. *Cyborg Selves: A Theological Anthropology of the Posthuman.* Farnham, UK: Ashgate.

Tilak, Bal Gangadhar. [1915] 1935. *Śrī Bhagavadgītā-Rahasya Or Karma-Yoga-Śāstra,* Volumes 1 and 2. Poona City: R.B. Tilak and Bombay: Bombay Vaibhav Press.

Tipler, Frank J. 1989. 'The Omega Point as *Eschaton*: Answers to Pannenberg's Questions for Scientists'. *Zygon: Journal of Religion and Science* 24(2): 217–253.

Tipler, Frank J. 1994. *The Physics of Immortality: Modern Cosmology, God and the Resurrection of the Dead.* New York: Anchor.

Tirosh-Samuelson, Hava. 2011. 'Engaging Transhumanism'. In *H+/-: Transhumanism and Its Critics,* edited by Gregory R. Hansell and William Grassie, pp. 19–52. Philadelphia, PA: Metanexus Institute.

Tirosh-Samuelson, Hava. 2012. 'Transhumanism as A Secular Faith'. *Zygon: Journal of Religion and Science* 47(4): 448–461.

Tirosh-Samuelson, Hava. 2013. 'Wrestling with Transhumanism'. In *The Transhumanist Reader: Classical and Contemporary Essays on the Science, Technology, and Philosophy of the Human Future,* edited by Max More and Natasha Vita-More, pp. 19–54. Malden, MA: Wiley-Blackwell.

Tirosh-Samuelson, Hava. 2018. 'In Pursuit of Perfection: The Misguided Transhumanist Vision'. *Theology and Science* 16(2): 200–222.

Torres, Ricard. 2013. 'QWERTY vs. Dvorak Efficiency: A Computational Approach'. Online. Available: https://pdfs.semanticscholar.org/3fff/a70a1a248b9 4654439f3758e5db83adc2904.pdf (accessed October 30, 2018).

Tripathi, Kapil. 2005. 'Recent Developments in Science & Technology'. *Dream 2047* 6(10): 18.

Trothen, Tracy J. and Calvin Mercer (eds.). *Religion and Human Enhancement: Death, Values and Morality*. New York: Palgrave.

Trovato, Gabriele, Loys de Saint Chamas, Masao Nishimura, Renato Paredes, Cesar Lucho, Alexander Huerta-Mercado, and Francisco Cuellar. 2019. 'Religion and Robots: Towards the Synthesis of Two Extremes'. *International Journal of Social Robotics*. Advance online access.

Trovato, Gabriele, Cesar Lucho, Alexander Huerta-Mercado, and Francisco Cuellar. 2018. 'Design Strategies for Representing the Divine in Robots'. *HRI 2018 Companion – Companion of the 2018 ACM/IEEE International Conference on Human-Robot Interaction*. IEEE Computer Society, Vol. Part F135192: 29–35.

Trovato, Gabriele, Cesar lucho, Alvaro Ramón, Renzo Ramirez, Laureano Rodriguez, and Francisco Cuellar. 2018. 'The Creation of SanTO: A Robot with 'Divine' Features'. *2018 15 International Conference on Ubiquitous Robots*. Institute of Electrical and Electronics Engineers, pp. 437–442.

Turner, Fred. 2006. *From Counterculture to Cyberculture: Stewart Brand, the Whole Earth Network, and the Rise of Digital Utopianism*. Chicago: University of Chicago Press.

Uberoi, Ptricia, Nandini Sundar, and Satish Deshpande (eds.). 2007. *Anthropology in the East: Founders of Indian Sociology and Anthropology*. Ranikhet: Permanent Black.

Ulam, Stanislaw. 1958. 'Tribute to John von Neumann'. *Bulletin of the American Mathematical Society* 64(3, Pt. 2): 1–49.

Underwood, Corinna. 2017. 'The Future of Artificial Intelligence According to Ben Goertzel'. *Techemergence*, December 9. Online. Available: https://www.techemergence.com/the-future-of-artificial-intelligence-according-to-ben-goertzel/ (accessed November 5, 2018).

Vahia, Mayank. 2015. 'Evaluating the Claims of Ancient Indian Achievements in Science'. *Current Science* 108(12): 2145–2148.

Vajpeyi, Ananya. *Righteous Republic: The Political Foundations of Modern India*. Cambridge, MA: Harvard University Press.

Van Dijck, José. 2014. 'Datafication, Dataism and Dataveillance: Big Data between Scientific Paradigm and Ideology'. *Surveillance & Society* 12(2): 197–208.

Van Otterlo, Martijn. 2014. 'Automated Experimentation in Walden 3.0: The Next Step in Profiling, Predicting, Control and Surveillance'. *Surveillance & Society* 12(2): 255–272.

Vance, Ashlee. 2012. 'The Ray Kurzweil Show, Now at the Googleplex'. *Business Insider* 4310: 55–56.

Varma, Vinayak. 2018. 'Omni8'. *Strange Worlds! Strange Times!: Amazing Sci-Fi Stories*, edited by Vinayak Varma, pp. 65–90. New Delhi: Speaking Tiger.

Vasavani, T.L. 1926. 'Asvamedha Or the World Sacrifice'. *The Kalpaka* 21(2): 49–56.

Velasquez-Manoff, Moises. 2020. 'The Brain Implants that Could Change Humanity'. *The New York Times*, August 28. Online. Available: https://www.nytimes.com/2020/08/28/opinion/sunday/brain-machine-artificial-intelligence.html (accessed August 31, 2020).

Vidal, Denis. 2007. 'Anthropomorphism or Sub-anthropomorphism? An Anthropological Approach to Gods and Robots'. *The Journal of the Royal Anthropological Institute* 13(4): 917–933.

Vigyan Prasar. n.d. Vigyan Prasar webpage. Online. Available: https://vigyanprasar.gov.in/ (accessed May 22, 2019).

Vinge, Vernor. [1981] 2001. *True Names*. In *True Names and the Opening of the Cyberspace Frontier*, edited by James Frenkel, pp. 239–330. New York: Tor Books.

Vinge, Vernor. [1993] 2003. 'Technological Singularity'. Online. Available: https://www.frc.ri.cmu.edu/~hpm/book98/com.ch1/vinge.singularity.html   (accessed April 2, 2019).

Visvanathan, Shiv. 1997. *A Carnival for Science: Essays on Science, Technology, and Development*. New Delhi: Oxford University Press.

Visvanathan, Shiv. 2017. 'Forward'. In *Alternative Futures: India Unshackled*, edited by Ashish Kothari and K.J. Joy. Self-published with Authors Upfront. Kindle edition (unpaginated).

Vita-More, Natasha. 2018. *Transhumanism: What Is It?* New Providence, NJ: Bowker.

Wachowski, Laurence and Andrew Wachowski. 1999. *The Matrix*. Los Angeles: Warner Bros.

Wadhawan, Vinod Kumar. 2005. 'Smart Structures and Materials'. *Resonance* 10 (11): 27–41.

Wadhawan, Vinod Kumar. 2007. 'Robots of the Future'. *Resonance* 12 (7): 61–78.

Wadhawan, Vinod Kumar. 2007. *Smart Structures: Blurring the Distinction between the Living and the Nonliving*. Oxford: Oxford University Press.

Wadhawan, Vinod Kumar. 2010. *Complexity Science: Tackling the Difficult Questions We Ask About Ourselves and About Our Universe*. Moldova: Lambert. PDF version downloaded from Academia.edu (accessed May 15, 2019).

Wakefield, Jane. 2019. 'Elon Musk Reveals Brain-Hacking Plans'. *BBC News*, July 17. Online. Available: https://www.bbc.com/news/technology-49004004 (accessed October 25, 2019).

Wallach, Wendell and Colin Allen. 2009. *Moral Machines: Teaching Robots Right from Wrong*. New York: Oxford University Press.

Warwick, Kevin. [1997] 2004. *March of the Machines: The Breakthrough in Artificial Intelligence*. Chicago: University of Illinois Press.

Warwick, Kevin. 2003. 'Cyborg Morals, Cyborg Values, Cyborg Ethics'. *Ethics and Information Technology* 5(3): 131–137.

Warwick, Kevin. 2002. *I, Cyborg*. London: Century.

Warwick, Kevin. 2020. 'Superhuman Enhancements via Implants: Beyond the Human Mind'. *Philosophies* 5(3): 1–10.

Watts, Pauline Moffitt. 1985. 'Prophecy and Discovery: On the Spiritual Origins of Christopher Columbus's "Enterprise of the Indies"'. *The American Historical Review* 90(1): 73–102.

Webb, Robert L. 1990. '"Apocalyptic": Observations on a Slippery Term'. *Journal of Near Eastern Studies* 49(2): 115–126.

Weizenbaum, Joseph. 1976. *Computer Power and Human Reason: From Judgment to Calculation*. San Francisco: W.H. Freeman and Company.

Wertheim, Margaret. 2000. *The Pearly Gates of Cyberspace*. London: Virago.

White, Andrew. [1896] 1923. *The History of the Warfare of Science with Theology in Christondom*. New York: D. Appleton and Company.

Wiener, Norbert. 1950. *The Human Use of Human Beings*. New York: Houghton Mifflin.

Wiener, Norbert. 1964. *God & Golem, Inc.: A Comment on Certain Points Where Cybernetics Impinges on Religion*. Cambridge, MA: The M.I.T. Press.

Wikipedia. 2019. '2.0 (film)'. Online. Available: https://en.wikipedia.org/wiki/2.0_ (film) (accessed May 21, 2019).

Wikipedia. 'Sapiens: A Brief History of Humankind'. Online. Available: https:// en.wikipedia.org/wiki/Sapiens:_A_Brief_History_of_Humankind (accessed May 27, 2019).

Wilczek, Frank. 2019. 'The Unity of Intelligence'. In Possible Minds: Twenty-Five Ways of Looking at AI, edited by John Brockman, pp. 64–75. New York: Penguin.

Wilson, Benjamin, Judy Hoffman, and Jamie Morgenstern. 2019. 'Predictive Inequity in Object Detection'. Online. Available: https://arxiv.org/pdf/1902.11097.pdf (accessed July 2, 2020).

Wolfram, Stephen. 2019. 'Artificial Intelligence and the Future of Civilization'. In Possible Minds: Twenty-Five Ways of Looking at AI, edited by John Brockman, pp. 266–284. New York: Penguin.

Young, George. 2012. The Russian Cosmists: The Esoteric Futurism of Nikolai Fedorov and His Followers. New York: Oxford University Press.

Young, William. 2019. 'Reverend Robot: Automation and Clergy'. Zygon: Journal of Religion and Science 54(2): 479–500.

Zargar, Arshad R. 2020. 'Privacy, Security Concerns as India Forces Virus-Tracing App on Millions'. CBS News, May 27. Online. Available: https://www.cbsnews.com/ news/coronavirus-india-contact-tracing-app-privacy-data-security-concerns-aarogya-setu-forced-on-millions/ (accessed July 8, 2020).

Zarsky, Tal Z. 2012. 'Automated Prediction: Perception, Law, and Policy'. Communications of the ACM 55(9): 33–35.

Zhang, Hai-Tian, Fan Zuo, Feiran Li, Henry Chan, Qiuyu Wu, Zhan Zhang, Badri Narayan, Koushik Ramadoss, Indranil Chakraborty, Gobinda Saha, Ganesh Kamath, Kaushik Roy, Hua Zhou, Alexander A Chubykin, Subramanian K.R.S. Sankaranarayan, Jong Hyun Choi, and Shriram Ramanathan. 2019. 'Perovskite Nickelates a Bio-Electronic Interfaces'. Nature Communications 10(1651): 1–7.

Złotkowski, Jakub, Ashraf Khalil, and Salam Abdallah. 2020. 'One Robot Doesn't Fit All: Aligning Social Robot Appearance and Job Suitability from a Middle Eastern Perspective'. AI and Society 35(2): 485–500.

Zuboff, Shoshana. 2015. 'Big Other: Surveillance Capitalism and the Prospects of an Information Civilization'. Journal of Information Technology 30(1): 75–89.

Index